Eureka!

The BIG book of science, poetry, coloring, note-taking, art, theorizing, sketching, music, graphing, inspiration, drawing, measuring, sculpture and creativity
(color the title if you would like)

by

Gary McCallister

THIS BOOK BELONGS TO:

(Leave blank if you really don't want
people to know who owns this book.)

Published by
Flaming Moth Productions
558 Casa Rio Ct.
Grand Junction, Colorado 81507

Library of Areyouserious?

WAFT* Edition
Cheap version

Executive editor: anonymous
Development Director: anonymous
Editor in Chief: anonymous
Creative Director: anonymous
Dover Design: anonymous
(you'll understand after you read the book)

*WAFT - Who asked for this? *

TABLE OF CONTENTS

(to be colored as desired)

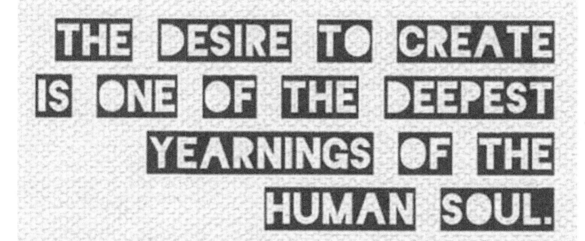

THE DESIRE TO CREATE IS ONE OF THE DEEPEST YEARNINGS OF THE HUMAN SOUL.

-DIETER F. UCHTDORF

Chapter 1

THIS IS A BOOK ABOUT CREATIVITY

Art, like morality, consists in drawing the line somewhere.
G. K. Chesterton

You may wonder why we need another book about creativity. I did. But having desperately sought creativity for a lot of my life, I realized that almost all the other books on creativity I have read, and they are legion, weren't very creative. There are many books approaching the subject of creativity from numerous angles. There are:

- Art books teaching artistic skills of various kinds.
- Psychology books about creative techniques for becoming creative.
- Science books about how the brain works in creative people.
- Business books are fostering creativity in a business.
- Education books about teaching creatively.
- Education books about fostering creativity in students.
- Craft books about creating crafts.
- And many others.

The problem with all these approaches isn't that they aren't very creative. Well, at least they never seemed to make me more creative. But the problem seems to be that the reader is never allowed to do any creating. They just read about it. I have worried about that a lot, but that probably just tells you something about me. My wife thinks it speaks volumes.

But then it struck me! I mean, like a real honest to goodness revelation, or inspiration, or something. Honestly, the thought was almost creative.

(You can color that if you are feeling creative.)

God took care of all that creating way back when. The details aren't exactly clear about how he did it, but apparently, whether you are an artist or scientist, there was a lot of stuff that was "without form and void". No one is sure what that was like because everyone since then has lived in a world that has form and, so far, hasn't been voided.

Another way of saying this is that matter, whatever that is, was "unorganized". So I guess God just organized it all, giving it form and avoided "voidness". I don't know why

6

He did this. Apparently, He just sort of likes to take a mess that hasn't any form or organization and trying to turn it into something special. My wife thinks he is still at it, in one form or another. She's waiting patiently for him to do something with me.

So, see? It's all done. Only took him a little while to accomplish it, too. I think it was on his day of rest that He thought of all the other things He could still do with this new thing. I think that's when it dawned on Him how much fun it was to create and that some of his creations might enjoy creating also. So after his day of rest he turned the whole project over to His creations to finish creating.

So, technically, you and I don't create anything. But we do get to re-create stuff from all the things that God created. And in doing so, we get to create our own world. I suppose it hasn't turned out so hot, but hey, we're beginners. In the process of re-creating with the stuff that God created we also get to create who we are and who we will become.

Of course, I think God doesn't expect too much out of the whole thing. Like I said, we're beginners. My understanding is that eventually He is going to hold a master class and show us how we might have done things better. I'm not sure how I feel about that. On one hand I hope He hangs my pictures on the refrigerator. On the other hand I suspect I will cringe when I see how immature they really are.

But God has arranged for further education in creativity if we are willing to learn. The way I understand it is that everyone gets to go on to the next grade because God stacked the panel of judges with His son and given him strict instructions to let everyone into the upper grade.

Okay, I think I'm rambling. The point is all those other books on creativity skipped the part about really creating anything. The best we can do is re-create with stuff that God has already created. The other books didn't have any . . .

RE-CREATION

(You can color that also, if you want.)

THIS IS A BOOK ABOUT CREATIVITY

The difference is that in this book you can create. Well, alright you can re-create. There are essays to read, pages to color, lined pages in which to write your thoughts and feelings, blank pages to draw or paint on, and poetry to lift your spirits. It's a lot of fun. In fact, this book is full of clean

RECREATION

(Be creative and color me)

More than 150 years ago George McDonald wrote a book entitled "The Diary of an Old Soul". In this book he wrote a seven-line poem for every day of the year.

Opposite every poem, however, was a blank page which he invited readers to compose their own poetry. In his dedication for the book he wrote this.

"Sweet friends receive my offering. You will find
Against each worded page a white page set: -
This is the mirror of each friendly mind
Reflecting that. In this book we are met.
Make it, dear hearts, of worth to you indeed: -
Let your white page be ground, my print the seed,
Growing to golden ears, that faith and hope shall feed."

I don't know that I know much about creativity. But it seems to me that this is a wise approach. And George McDonald is a man I think I could emulate profitably. So I have taken a similar approach. Not only will you find a blank page, but a lined page for thoughtful writing, essays, poetry, scientific questions, and quotes from famous, occasionally infamous, creative individuals to encourage your own creativity. But don't worry. There are not 365 of them.

Through a lifetime of scientific research and artistic pursuits I have had a lot of fun. My hope is that when you have finished this book you will feel as if you have experienced some refreshing recreation, and perhaps you will feel a little more creative in your own life.

So turn the page and

DISCOVER

A little more about creativity.

"The chief enemy of creativity is good sense." — Pablo Picasso

(you can color this, also. After all, it is Picasso.)

Chapter 2

SEEING THE DARK

"Imagination is everything. It is the preview
of life's coming attractions."
Albert Einstein

Humans can see the "dark". They just can't see "in the dark". This seems strange to me. Of course, we can see light, and we can see in the light. But how do we see dark? Especially, how do we see dark when it's light?

Why is it dark at night? Oh sure, the sun is on the other side of the earth. But If the universe is infinitely large, holds an infinite number of stars, is homogeneous and static, the night sky should be luminous. Wherever you looked there would be another star to light up the sky. While the number of stars visible is large, they surely don't light up a luminous sky.

Honestly, don't people who come up with this kind of stuff have a job to do or something? When I sit around and make up stories that are patently not true, I'm called a liar. Personally, I prefer using the term historical fiction.

Obviously, the night sky is dark. All you must do is look. So, one or more of the above assumptions must be false.
- the universe is infinitely large.
- holds an infinite number of stars.
- is homogeneous and static.

This quandary is known as a paradox, where logical conclusions are contradictory of fact. The "dark sky paradox" is commonly attributed to the German amateur astronomer, Heinrich Wilhelm Olbers, who described it in 1823. It is often called the "Olbers' Paradox" in his honor. That is if it is an honor to come up with stuff that is paradoxical.

My wife could argue that my life is a paradox because my logical conclusions are usually inconsistent with the facts. You may wonder why she stays. I do. I think the truth is that she is secretly an adrenalin junky.

Interestingly, the first person to suggest a solution to Olbers' paradox was Edgar Allan Poe. Oh, yes, the poet! And you probably thought science wasn't creative. He pointed out that either some stars must be so distant that their light has not reached us, or that some stars were still moving away from us at a high rate of speed.

Of course, he didn't know about the Big Bang or the "speed of light" in 1849. So his was an impressive observation. Poe published his proposed solution in a book-length work entitled Eureka in 1849, shortly before he died. It is one of his lesser-known works, probably because it was his only venture into cosmology and astrophysics. That's seems strange though. Cosmology is so often a big seller in the book world.

Anyway, he pointed out that since the universe is finitely old and the speed of light is finite, only finitely many stars can be observed within a given volume of space visible from Earth. The density of stars within this finite volume is sufficiently low that any line of sight from Earth is unlikely to reach a star. Thus either light from some stars have not yet reached us, or some stars are moving away from us faster than the speed of light.

The Big Bang theory introduces a new form of the paradox. Following the Big Bang, assuming there was one, the universe was basically filled with dust. As the dust settled and the universe became transparent, all points of the local sky in that era would have been of comparable brightness to the surface of the Sun, making the entire sky light.

Hello! It's dark out at night. Darkness, then, is claimed to be evidence that either the universe is not infinitely large, doesn't have an infinite number of stars, or is not homogeneous and static. Well, that narrows it down quite a bit.

Of course, we believe now that the universe is not static but expanding because of the initial explosion. However, knowing this does not change the paradox of a dark night sky significantly. Numerous explanations have been proposed, although unless you are one of the proposers, none of them are completely satisfactory. It is a little disconcerting when we discover that science can't explain why it's dark at night!

Poe's explanation may not be right. Scientists like to point out that his book, Eureka, does indeed evoke some modern scientific thought, but it does so in a very blurry way. Admittedly, considering the source, Poe's ideas may have been drug induced.

Personally, I think it is simply professional jealousy. I mean, for Pete's sake, Poe was a poet, not a scientist. But maybe cosmologists could use a little help from poetry now. Their present theories are only about that useful.

However, this is an important concept to start this book because it illustrates the fascinating overlap of creativity in science and art. Oh, I know most folks don't think science is creative. They think you simply memorize a lot of stuff and you become a scientist. But it turns out that there is more creativity in science than is popularly understood, and more science in art than is appreciated.

However, the two camps have tended to separate in recent years due to specialization and most modern poets aren't thinking much about "Olbers' Paradox", and most modern scientists aren't reading long, prose poems like Eureka.

Speaking of Eureka, the very term for sudden insight and discovery often associated with the arts was coined by a scientist, Archimedes. You see, most of what we think we know we accept from someone else telling us it is true. But when we have our own real-life experiences, we often have sudden discoveries that are life changing. Archimedes had one of these. I'll tell you the story in the next chapter.

"You can't use up creativity. The more you use, the more you have."

− Maya Angelou

SEEING THE DARK: a poem

Can you dance and make it rain?
Command the soil to produce grain?
Hold the sun in place by your hand?
Clear the desert of the sand?

Can you stop a falling star?
Heal a wound without a scar?
Put a halt to the oceans tide?
Give life to one who's died?

Sorry. Creation's already done.
The Father did it with his son.
But it still seems to be man's fate
To forever re-create.

SEEING THE DARK: science

Is there a relationship between scientific principles and artistic ideas?

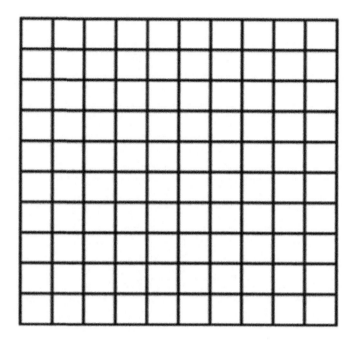

How could you illustrate a scientific idea with artistic expression?

How could you express an artistic idea scientifically?

How would one illustrate, or scientifically measure "darkness"?

SEEING THE DARK: coloring page

Black is said to be the absorbance of all wavelengths of light. White is said to be the reflection of all the wavelengths of light. So what is absorbing all the wavelengths of light at night that reflects all the wave lengths of light in the day?

SEEING THE DARK: sketch - draw - paint - paste - fold - cut - etc.

Chapter 3

CREATIVITY AND ARCHIMIDES

*Creativity involves breaking out of established patterns
in order to look at things in a different way.*
Edward de Bono

How can we ever know what's "true"? Science is supposed to take care of that. Except it doesn't when the truth may be about non-material things such as God, love, patriotism, or values. Science isn't always even helpful when dealing with the material world because we depend on scientists to figure things out. And scientists are, well, people. People make mistakes, misinterpret data, and sometimes even lie.

Sorry! Lie is a harsh word and it isn't acceptable to use harsh words anymore, even if those words are the truth. I probably should have called a lie "fake data." In addition, most of us rely on the media to report what's happening in science, but all too often the media adds an entire layer of lies, I mean, misunderstandings.

In other words, most of us rely on other people to tell us what is "true". There is nothing wrong with that. We can't all verify everything for ourselves. I believe Africa exists although I haven't seen it. My wife tells me what to do all the time, and I am extremely obedient. (She'll probably edit that out.) But I do what she says because I trust her.

Another way of determining truth is by our own experiences. You might say experience is a form of science. Or maybe science is a special form of experience. Either way, I think most people have had experience with people who are not experienced, and it is often not good. That is why employers hire people with experience.

There is an interesting story about learning from experience. The story goes that Archimedes, a mathematician who lived two centuries before Christ, noticed his bathwater would rise when he got into the bathtub. He then realized that the volume of water he displaced with his body had to be the volume of his body. He is supposed to have jumped from his tub shouting "Eureka! Eureka!" as he ran down the street without his clothes on.

You might assume that he was just another strange scientist to be so excited as to forget his clothes after discovering the displacement of water. However, his excitement was because he had just discovered how to save King Hieron's dilemma concerning how to tell the truth. He assumed he would be rewarded royally for his discovery, and that was something to get excited about, and possible some nice new clothes.

See, the King had given a large amount of gold to a jeweler named Dionthenes who was to make his crown. After the crown was delivered, it dawned on the King that Dionthenes might have used a baser metal for the interior of the crown, used the gold to plate the outside, and kept the rest of the gold for a profit.

The King couldn't exactly cut the crown in two to see, so he asked Archimedes how he could determine the truth. Archimedes reasoned that baser metals, like silver, were less dense than gold and would take up more space. Knowing the amount of gold, the King had given the jeweler, Archimedes could immerse that same amount of gold in water and note the displacement. Then he could place the crown in water. If the interior of the crown was made of a less dense metal, it would have greater volume and displace more water.

It turned out that the inside of the crown was made of base metal. History has not recorded what happened to Dionthenes, although we can assume it was not good. However, the word "eureka" was permanently recorded, probably due to the shock Archimedes caused when running around naked. The word "heuristic" was derived from the word eureka and means discerning the truth from experiences. Discerning truth from personal experience is at the very heart of art and science efforts to be creative.

Sometimes it's hard to know the truth. Science is supposed to take care of that. But it seems like I never have all the information I need to guarantee that I know the truth. So, I must go with experience.

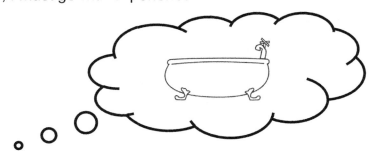

CREATIVITY AND ARCHIMIDES: A poem

An owl outside my window
For seven nights in a row
And other sounds of the night
Like the tiny scream of a rabbit's plight
As I paint by firelight

Painted by the dark of the moon
Images of the afternoon
When I planted before last frost
To lengthen the time of Pentecost
And prayed not all would be lost

The nature of creative fire
Does it burn or inspire?
My painting illuminates
From the fire in the grate
Did I discover, or create?

CREATIVITY AND ARCHIMIDES: thoughts

CREATIVITY AND ARCHIMIDES: science

How could one illustrate density?

Does the concept of density inspire an image or metaphor?

DENSITY

CHANGE IN WATER LEVEL

What is the science behind the concept of density as it is applied to art? Music?

CREATIVITY AND ARCHIMIDES: coloring page

Chlorophyll in plants make energy for the plant from sunlight. There fore plants are benefitted by receiving as much light as possible. In a tree, the plant needs to maximize the amount of light it can receive, so there is a competition for light. The tree needs to grow so the leaf surface is as thick and "dense" as possible. What is the most efficient shape for increasing surface area? What is the greatest shape for obtaining the thickest density?

PS Like a lot of o scientists, Archimedes was a bit of a fanatic. He is said to have died in Sicily when the Romans captured it following a siege that ended in either 212 or 211 BCE. One story told about Archimedes' death is that he was killed by a Roman soldier after he refused to leave his mathematical work. The last words attributed to Archimedes are "Do not disturb my circles", a reference to the circles in the mathematical drawing that he was supposedly studying when disturbed by the Roman soldier. This quote is often given in Latin as "Noli turbare circulos meos," but there is no reliable evidence that Archimedes uttered these words.

CREATIVITY AND ARCHIMIDES: sketch - draw - paint - paste – fold – cut - etc.

Chapter 4

SPONTANEOUS ORDER

"Doors are for people with no imagination."
Derek Landy

It's strange that anything ever "gets organized". The laws of physics say that everything should become "less organized" over time. When I look at my office, I believe in physics. Physicists call this disorderly state "entropy". I like to call my office "entropic" because it sounds better than "chaotic". My wife just calls it a mess.

I have tried to explain to her that entropy, sometimes envisioned as randomness or disorder, applies to any enclosed system. Yes, I could spend energy and reduce the randomness in my office, but it would necessarily increase the randomness in the rest of the house. You'd think she would want to keep entropy somewhat isolated.

However, one of the great mysteries of modern science is "self-organization." Why do some things, on the edge of entropy, suddenly and seemingly self-organize and become more orderly? For example, there are well-documented reports of fireflies blinking on and off in unison over extremely large areas of land. How does this happen? This phenomenon develops faster than a dance floor full of people responding to the bass beat. Besides, there are usually no basses present to help the fireflies keep time.

My wife thinks that my problem is that my mind isn't organized, let alone "self-organized". I don't feel too badly about that though. No one knows what a mind is, so I guess no one knows how it is organized either. I don't think humans should take credit for organizing something they have no idea how it even works, lets alone how to organize it.

I guess teachers think that they are helping to organize student minds. But I don't think they really know what they are doing. I never did, but I still took credit for doing it. It is a little disconcerting when I recall that most of my world was built by people who didn't have near as much teaching as I have received. I can't wait to see how wonderful the world will be after my students grow up after all my mind organizing.

People always think they are self-made men or women when in fact they had very little to do with what they are. Their parents took care of a lot of it, but even they didn't know what they were doing. In fact, I believe most children are conceived in a thoughtless passion. Then the parents spend the rest of their lives trying to make up for their thoughtlessness.

I also think a lot of our minds are "God-organized", though some say that isn't very scientific. However, it's a scientific as saying we are self-organized when we don't really know how it's done at all. It's hard to tease apart, biology, God, and experience in the forming of a "self-organized" mind.

Cells from heart tissue, grown in an artificial-culture medium, all begin to pulse at a very early stage. If they are not touching, they each pulse at their own rates. However, after two cells grow together, they begin to pulse in unison. The mechanism of pulsing heart cells is well recognized in medicine, but I have never seen a good explanation for how the two regulate to the same rate.

A student and I once did some experiments on the growth of artificial membranes. Certain chemicals, when combined in the proper concentrations, form a

23

film in the fluid. This film separates the fluid into different sides of the membrane, and this separation alters which molecules are available to react on both sides.

As the membrane grows, the smaller molecules can pass through it and tend to do so by random energy. However, the larger molecules become trapped on one side of the membrane. Because the large molecules cannot move, they change the concentration on their side of the membrane and attract more small molecules. The membrane then swells into a plant-like structure in the beaker.

The change in concentrations of different molecules favors certain reactions over others. Further, the reaction that forms the membrane holds certain molecules in position, again favoring certain chemical reactions over others. These favored reactions continue the growth.

The energy that causes entropy causes an increase in structure and order in this case. Uniform order becomes visible where the prediction would normally be entropic disorder. This phenomenon of self-organization, sometimes called synchrony, was a surprise in the late twentieth century. It is still being explored today.

Are forms of art an example of spontaneous organization. You know, like there is this jumble of ideas going around in someone's head and suddenly it all takes form and becomes a symphony or sculpture. Even if the execution is methodical, is the decision about what to do spontaneous? Why doesn't my mental energy ever lead to art. My mental energy always ends up in entropy.

I have been trying to build a computer model that would mimic membrane organizing activity. The ability to understand how the energy that causes entropy can also create order would be a huge discovery on the part of science. It would allow humans to create order from universal functions that have been assumed to cause only entropy and chaos.

Personally, I have less-ambitious goals for my model. I am hoping to discover the exact conditions required for my office to self-organize like the artificial membranes. I have recently made some progress, by adding further clutter, to get the exact concentration of various components of musical instruments, books, papers, electronic parts, and projects in hopes of kick-starting the organization.

From there it should be easy. The one thing I am worried about is that, according to Newtonian physics, a decrease in entropy in one place leads to an increase in entropy elsewhere. This could get tricky. The other concern is that our artificial membranes don't last long. Within twenty-four hours or less, depending on agitation, they usually collapse and become disorderly again. My wife will keep you posted.

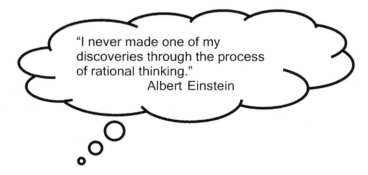

"I never made one of my discoveries through the process of rational thinking."
                          Albert Einstein

SPONTANEOUS ORDER: a poem

Thundering winds and lightening flash
Pounding rain beats down the grass
Rivulets stream across the rocks
Beneath spinning weathercocks
Gathering mud and life's debris
Gather together for their march to the sea

Flooding streams become swirling floods
Eroded red soil make it look like blood
Then cascading over rock and wall
It becomes a thundering waterfall
Boiling boulders from the ground
Ancient massive trees are downed

At last in single tributaries wide
The water slows its violent tide
And flows fast, wide and deep
Spreading arms with a mighty sweep
A Until it seeps under every door
And spreads its way across the floor

But in just a little way downstream
In the bank there's a tiny seam
A row of rock where there was none
Creates just the smallest diversion
Through which a small stream flows
To feed the fields where good things grow

From the violence of nature's plan
Beauty created from the mind of man

SPONTANEOUS ORDER: science

Do the rules of entropy apply to artistic production? For example:
- Since nature follows the rules of random distribution of energy do artists also follow this rule?
- Does an increase in order within a work of art decrease order someplace else?
- How can one portray randomness in a work of art? (Obviously it will be different in various media.)
- Can there be art without order?
- Is order in scientific theorems and ideas beautiful?
- Is science and math to symmetrical to be artistic?
- How many ways can three, six-stud Legos be connected?

Create an illustration of entropy in the matrix below.

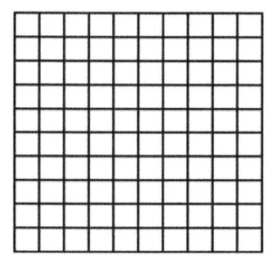

- How many ways can three Lego's that are connected, connect to three other connected Lego's? Does that image inspire a design?

SPONTANEOUS ORDER: coloring page

Can disorder ever be beautiful? If we see something that is disordered, like the sky at sunset, and it seems beautiful, does that mean that there is an order to it and we just can't see it? What is order? What is disorder?

SPONTANEOUS ORDER: sketch - draw - paint - paste - fold - cut - etc.

Chapter 5

ART AND THE CREATION OF SCIENCE

*"A rock pile ceases to be a rock pile the moment a single man
contemplates it, bearing within him the image of a cathedral."*
Antoine de Saint-Exupery

Abstractions are things you can't hold in your hand. Interestingly, science usually starts out with material things you can hold in your hand and ends up with abstract ideas. On the other hand, art often starts out with an abstract idea and ends up with a material object you can hold in your hand. Either way, we live in a material world.

The best way to learn about material things is to have experiences with materials, actual physical experiences. That is why scientists have labs and artists have studios, I guess.

A few years ago, we were doing a lab at the university using live frogs with upper-division students. Just to clarify, the experiment was supposed to be done on the frogs, by the students. After demonstrating the procedures of the lab exercise, I turned the students loose to get their own frogs. Soon there were loose frogs all over the lab with students on the floor hollering and chasing them. It was very amusing and entertaining.

Then it dawned on me that these upper-division biology students had never handled a live frog before and were not sure how to hold them. To their credit, they quickly learned a skill that probably should have been mastered when they were five or six years old playing outside in a ditch. They might have learned how to catch and hold frogs if their kindergarten classes had been allowed to play outside more.

Since that time in the lab, it has become clear to me that many children today live in an "experiential desert". Having this happen is less likely here in our western valley because we live so close to nature and opportunities. Unless the health department doesn't let you . . . But children in other parts of the world, and especially major cities, have very limited opportunities for rich and varied physical experiences. At best, schools may inhibit such opportunities simply because of their isolation, basic structures, and need for a safe atmosphere.

There is a great misunderstanding, in general, about science and art in the public schools. Many people believe science is essential and important because of its contributions to our lives. Some believe the arts are less important because they don't appear to directly further economic prosperity. Of course, this is obviously not true when we take time to examines the economic successes of the various artistic media.

Another contribution that the arts make is even less recognized. The arts can provide opportunities for physical experiences with the material world that are lacking in much of urban and suburban life and schools. Therefore the "Take Note" initiative announced for school district 51 providing musical experiences for public school children is significant.

When music is experienced by children, they develop concepts concerning degree, scales, wavelengths, movement, whole numbers and fractions that will later make concepts of number lines, analog and digital systems, spreadsheets, spatial arrangements, formulas and math relationships much easier to grasp. Children, who have not been free to run, can move freely about the keyboard. Children who have not had rich experiences with high and low, fast and slow, loud and soft, can relate to temperature, spectra, decibels, and periodic charts with greater ease.

Music education, and the arts in general, do not necessarily produce more musicians and artists. But experiences with the material world do create more scientists, engineers, accountants, technicians, mathematicians, and even businessmen. This is evident by the large number of technically skilled people who have music experiences in their background and who often continue to pursue these activities for pleasure.

Even older people can benefit from increased and varied activities in the arts. My wife has repeatedly suggested that I need to have more and varied physical experiences around the house. She says playing the guitar isn't exactly what she has in mind, even though I point out to her that I am still not very good.

And science is a bit of a Johnny-Come-Lately to the intellectual experience of humanity. There is much disagreement about exactly when science began, but generally scientific thinking has only been around for several hundred years. Humanity survived a very long time without it, although there is no question that science has greatly improved the human experience, gifting us with better nutrition like Twinkies, health practices like abortion, energy and nuclear bombs, and greater wisdom like . . . Well, you know what I mean. I won't even get into sciences contribution to plastics, pollution, climate change, environmental degradation and other sources of high paying jobs.

But it is extremely unlikely that science could have ever come into existence without the thousands of years of practical and artistic skill developed through mankind's early history. Skilled tradesmen and women passed down knowledge of artistic skills and practical manufacture without which science is not likely to have developed.

31

ART AND THE CREATION OF SCIENCE: a poem

WITHOUT AND WITHIN
I tried to change the system from within
Abstractions until my mind grew numb
Formulas and verbs again and again
For twenty long years of boredom

Then I saw all the world without
Felt the sensations on my skin
I heard sounds without a doubt
And "without" changed the system within

BEZALEL: another poem

Working In the shadow of God
Chief artisan of the tabernacle
Carver of wood
Joiner of frames
Weaver of textiles
Stitchery of thread
Master of metals
Guilder of gold
Sculptor of silver
Precision and art
Material knowledge
Craftsman skill
Without centuries of knowledge and skill
Could science have been created

ART AND THE CREATION OF SCIENCE: thoughts

ART AND THE CREATION OF SCIENCE: science

How is written music like a spreadsheet?

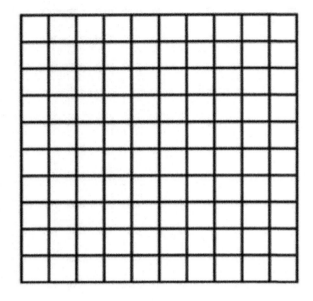

How is a painting like a spreadsheet? A sculpture?

How is a graph like a work of art?

Cn you create a three dimensional graph?

ART AND THE CREATION OF SCIENCE: coloring page

The periodic chart of the elements is thought to be beautiful to chemists. The layout and logic help them to understand relationships and make predictions. How could this diagram be colored so that it was a code that could be predicted? Or maybe just do it for fun and beauty.

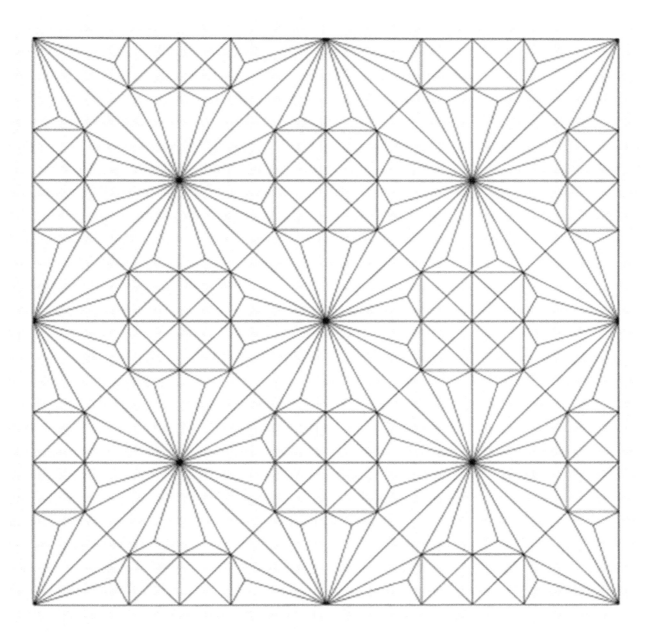

ART AND THE CREATION OF SCIENCE: sketch, art - draw - paint - paste - fold - etc.

Chapter 6

SCIENCE AND THE CREATION OF ART

"Their 'magic' is Art, delivered from many of its human limitations."
J.R.R. Tolkien, The Silmarillion

I'm probably the only scientist alive who doesn't really understand what science is. I don't know that for sure because I don't know all the scientists alive today. I assume a lot of them know what they are doing. I just never did.

I know all the stuff scientists say that science "is". The thing is most of the early, big discoveries of science were made by people who didn't have any clue about scientific method, tests of significance, or the national Science Foundation. I have a sneaking suspicion that chemistry was born from people having fun playing with fire and explosive materials. They were probably children. We don't allow that anymore.

Most early scientists were amateurs pursuing some hobby or project that had little, or nothing, to do with making a living. They are the ones who developed experimentation and observational techniques and made hypotheses before hypotheses were cool. There are still a lot of these kinds of people around. It's just that the real scientists usually don't listen to them.

Nowadays you must go to school to be a scientist, and meet standards like, "defining problems in terms of criteria and constraints, generating and evaluating multiple solutions, building and testing prototypes, and optimizing." That sure sounds a lot more impressive than what I ever did.

Mostly I just read a lot of stuff and thought about things I had read. You can't think about thinking without thinking about something. So thinking almost always ended up with me playing around with something. I did learn in school that if I played around carefully, I could call it an experiment and could get paid for it.

Sometimes I wasn't so much thinking as imagining. That's probably where I went wrong . . . Is there a difference between thinking and imagining? I think there must be, but I can't imagine what it is.

Anyway, thinking sometimes led to planning experiments. Of course, that led to doing the experiment. The truth is that "experimenting" is another name scientist use for playing around. I'll probably be disbarred by the Scientific Bar Association for letting that little secret out. I'll claim whistle blower status.

I think I did discover something interesting out of all the playing around with my experimentations. I learned people always want to show others what they have been playing around with. Maybe not right away, but eventually we all want to share what we have been doing. It doesn't matter if it is an experiment or a poem, a symphony or a computer program, we crave feedback.

You'd think we would learn. Showing people what you are doing always leads to trouble. If you've been thinking about something, I guarantee that there is someone out there thinking about it differently. They are always more than happy to tell you about it and show you what they have been doing that proves you are doing it completely wrong.

All this will cause you to go home and thoughtfully reflect about these differences. "Thoughtfully reflect" is scientific code for "brooding and feeling depressed." This sometimes leads to sad unfortunate outcomes. However, if you are a true scientist, from this brooding and depression you might just think of a new way of playing around and start another "experiment"!

So playing around generally begets more playing around. If you are a little lucky, and persistent, you might be able to be a scientist for a very long time like this. You must remain optimistic though. A lot of older scientists become cynical. I have avoided this problem although I sometimes slip into "failed optimism" stages. But I am never cynical.

My friend Gary Stager came up with an acronym for the "scientific method" that he called TMI. It never really took off, though, because most people thought it meant "Too Much Information" which is what most people associate with their science classes in school.

Of course, it really stood for "Think, Make, Improve". Stager is a computer scientist, and he saw that as the process for developing software. He liked it because it was a never-ending process that could be applied again and again forever. In computer science, this calling forth the same process is called "iteration" and is fundamental to programming.

However, because TMI never ends, a new problem is created. It never ends! Look at the Apple I-Phone, version 10, Microsoft Word 10, Beethoven's' Fifth Symphony. At least Beethoven knew to stop at nine . . . I told Gary that he was all wrong and that it should be IEI: imagine, experiment, improve. He told that sounded too much like E-I-E-I-O from Old MacDonald had a farm. Maybe IEP - imagine experiment, publish - would be better although most of my playing around, I mean experimenting, was never publishable.

Science has now provided techniques and ideas that supply the modern artist with new tools and new perspectives. So artists imagine, make, and show: IES? More data! Everyone always needs more data. That means everyone needs more money. Wait! What if it's the need for more money that generates the need for more data?

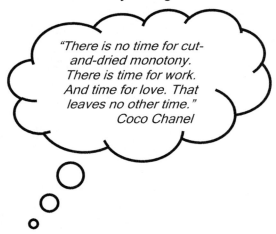

"There is no time for cut-and-dried monotony. There is time for work. And time for love. That leaves no other time."
Coco Chanel

SCIENCE AND THE CREATION OF ART: poem

CHEMILUMINESCENCE
(a poem by Gary McCallister)
They make their light by chemiluminescence
That alone explains their flashing brilliance
And explains their fleeting transience
They twinkle somewhat like a star
A flash of light like pulsing radar
But that's no longer where they are

FIREFLIES IN THE GARDEN
(a poem by Robert Frost)
Here come real stars to fill the upper skies
And here on earth come emulating flies
Although they never equal stars in size
(And they were never really stars at heart)
Achieve at times a very star-like start
Only, of course, they can't sustain the part

SCIENCE AND THE CREATION OF ART: thoughts

_____

_____

_____

_____

_____

_____

_____

_____

_____

_____

_____

_____

_____

_____

_____

_____

_____

_____

SCIENCE AND THE CREATION OF ART: coloring page

Can you turn the periodic chart of the elements into a work of art?

PERIODIC TABLE OF ELEMENTS

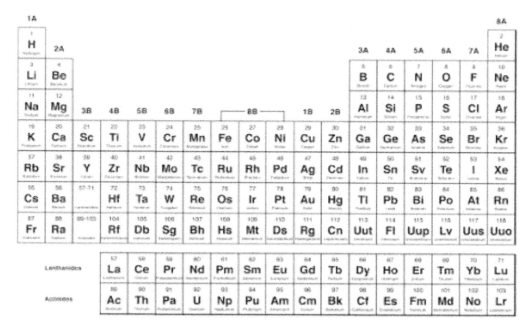

www.AllFreePrintable.com

41

SCIENCE AND THE CREATION OF ART: sketch - art - draw - paint - paste - fold - etc.

Chapter 7

BY THE WATERS OF BABYLON

"By the rivers of Babylon, there we sat down, yea,
we wept, when we remembered Zion."
Psalms 137

One of the comforting things about science is that the world makes sense. We may not always understand what is going on, but there are rules, that can be followed faithfully if we simply figure them out. And there's no cheating. Oh, I suppose there are cheating scientists. We're only human. But the material world doesn't cheat. You fall if you step off a cliff - regardless of who you are. It's not like humans. It's hard to trust almost anything about humans. We are inconsistent, mercurial, biased, and ultimately fascinating.

This thought came to my attention as I was thinking about distance and direction while traveling a lonely stretch of highway in southern Idaho. If you've driven there, you know why. There's not a lot of anything else to think about if your wife falls asleep and stops reading to you.

The compass in the car wandered back and forth from north to south and generally west. Does anyone else have a compass in their car anymore? Mine is a very old car. I don't know what degree northwest is on a compass; maybe thirty degrees. Suddenly, I wondered why we divide the circumference of a circle into 360 degrees.

When I was young, I learned that we measure the world in things like pounds, feet, quarts, and teaspoons. Then I was taught in school that these measurements were imprecise and old fashioned. The better method of measurement was the metric system: using grams, meters, and other units using a base of ten.

So why is the circle not divided into 100 units? A quarter turn should be twenty-five degrees, not ninety. The time of "six-thirty" ought to be six-fifty. Due west would be seventy-five degrees. How in the world did we get to three-hundred-sixty degrees in a circle?

I tend to think that I am at the pinnacle of social advancement, far smarter than all the people who have lived before. So, it was a terrible shock for me to discover that our 360-degree circle is based on a 2000-year-old mathematical system used by the Babylonians!

The Babylonians were a rich, but idolatrous, people who destroyed Jerusalem and enslaved the Jewish people many years ago. As such, the term Babylon continues to be a symbol of depravity. What? They were mathematicians? That illustrates that mathematicians have a natural tendency for torturing people.

Apparently, they developed a place-value system like the base ten of the metric system, only theirs used the base sixty. They had a different representation for all numbers up to fifty-nine, and then they started a new column representing one lot of sixty, plus the next number.
The number sixty can be divided in many ways, so all they had to do was keep track of how many groups of sixty they had, plus how many were left over.

That may seem awkward to us, because we aren't used to counting on our fingers. In fact, I was expressly forbidden to use my gingers for counting when I was in school. That always made me wonder what my fingers were for if it wasn't counting. Later I learned how to play the guitar and understand that fingers had more important uses. Still, seems like the could be used for counting if a person wanted.

However, our objection to using fingers to count may stem from our lack of knowledge of anatomy. Two thousand years ago the Babylonians knew that each finger had three bones, not counting the thumb. There are four fingers on each hand, so there are twelve different bones. Twelve divides evenly into sixty. The left hand was used to count to twelve counting each bone as one. The right hand was used to keep track of how many twelves are counted. With the two hands, they could count to five groups of twelve, or sixty.

The Babylonians divided the circle into six groups of sixty, or 360 units. So, we modern, scientifically advanced, people still have sixty seconds in a minute, sixty minutes in an hour, and 360 degrees in a circle because of the way unrighteous slaveholders did their math 2000 years ago.

The distance across southern Idaho is the same regardless of which unit we use to describe it. The fact that humans can't make up their minds about which units to use to describe the distance is just typical of fascinating humanity.

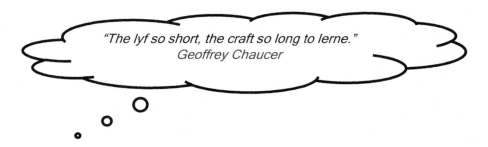

"The lyf so short, the craft so long to lerne."
Geoffrey Chaucer

BY THE WATERS OF BABYLON: poem

A SONG OF SORROW (Psalms 137:1-6)

"By the rivers of Babylon,
there we sat down,
yea, we wept,
when we remembered Zion.

We hanged our harps
upon the willows
in the midst thereof.

For there they that carried us away captive
required of us a song;
they that wasted us
required of us mirth,
saying, Sing us one of the songs of Zion.

How shall we sing the Lord's song
in a strange land?
If I forget thee, O Jerusalem,
let my right hand
forget her cunning.
If I do not remember thee,
let my tongue
cleave to the roof of my mouth . . ."

BY THE WATERS OF BABYLON: science

How would a drawing based on the number twelve appear different than one based on the number ten?

How many degrees would there be in a circle if we divided the circle by the number ten?

How could the number twelve be illustrated to demonstrate the evils of Babylon for the Jewish people?

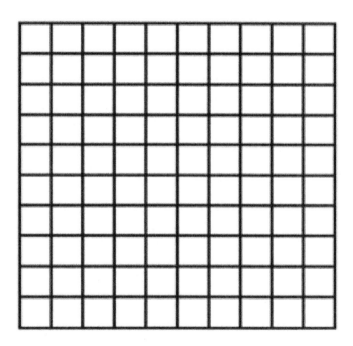

How could one portray the number system with a base 60 in art?

Did you realize that there are twelve tones in an octave of musical sound?

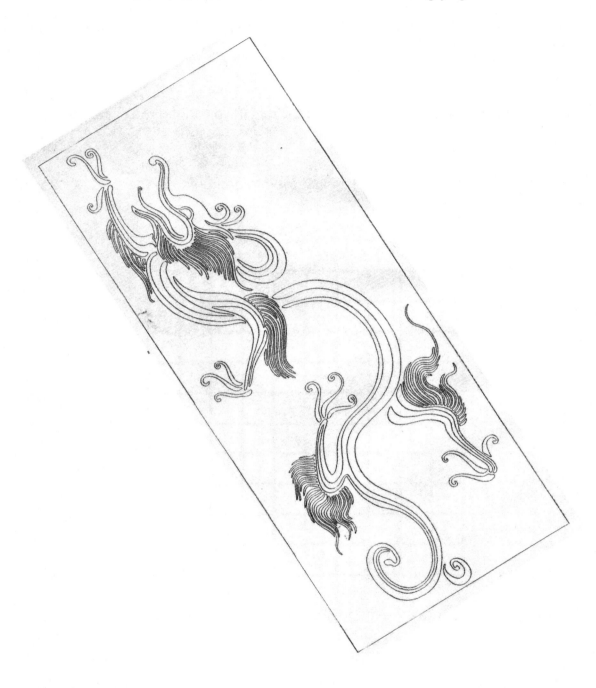

BY THE WATERS OF BABYLON: sketch -art - draw - paint - paste - fold - etc.

Chapter 8

PATTERNS

*"My life seemed to be a series of events and accidents.
Yet when I look back, I see a pattern."*
Benoît B. Mandelbrot

If I must have male-pattern baldness, I would like it to be in zigzags. Then I would at least look dangerous. My wife says it would just look ridiculous. I thought about wearing camouflage to look dangerous, but I couldn't find any. One of the disheartening things about men growing old is that, at some point, no one sees you as much of a danger anymore. I don't think women are ever perceived as less dangerous.

Categorizing people is part of our human ability to recognize patterns. When we see a person who we think shares some characteristics with another person we tend to create in our mind a category. Dangerous people are a category. Often categories overlap. For example, dangerous people and women are two different categories. However, you can be dangerous and not be a woman, but all women are dangerous. It's the pattern that matters.

Identifying patterns is what scientists do. If we can identify the pattern of events in time or space, we can then predict the event. If we can discern what causes the pattern, we can work to control it. Identifying patterns can be more difficult than you might think because patterns can occur over time, or across space, and sometimes both simultaneously. In addition, patterns can occur over extremely small, invisible-to-the-naked-eye scales like atoms and cells, or large, difficult-to-comprehend scales such as geologic time or endless space.

Patterns can sometimes be so complex that we can't even see them. That doesn't mean there isn't a pattern, only that we haven't found it yet. That raises the question as to whether there is such a thing as randomness. Maybe there are only patterns on a scale we can't discern or explain. Even when we see a pattern, we may not know it's cause.

For example, male-pattern baldness is the loss of hair in men as they age, usually evidenced in receding hairlines and loss of hair on the crown. But the loss of hair isn't due to loss of hair follicles or, even, hair. It's that the hairs themselves are so thin as to be invisible. And, each hair is so thin that it is easily broken off. I tried to tell the people at the driver's license bureau that my hair was there it was just flesh colored. They said there was such a category. Scientists have discovered that a protein present in the scalps of some men, derived from genetic factors, binds to the hair follicle and inhibits proper hair formation.

It can be curiously difficult to tell biological patterns from simple physical patterns. I think most people associate some pattern with living things, although that pattern can be a little difficult to put into words. We assume some kind of regularity, like bilateral symmetry or tree branching, but not a geometric regularity like a crystal.

ALH84001 is a meteorite discovered in Allan Hills, Antarctica in 1984. It is thought to have originated from Mars four billion years ago when a meteor blasted it loose. It apparently arrived on earth about 13,000 years ago. I don't know where it was in the interim, sort of like teenagers between the time they leave home and the time they arrive home several hours' past curfew.

The meteorite became famous when some scientists claimed to have found evidence for microscopic fossils of Martian bacteria in it. Eventually, the observations were explained using purely physical explanations. However, the imminent scientist, President Bill Clinton, gave a speech about the discovery, though he knew nothing about science. This caused much attention, so the discovery of this meteorite is considered the historical birth of the field of astrobiology.

In other words, astrobiology became a new category of science even though the pattern didn't fit. Of course, since 1984, no one has found any evidence of astrobiology. But there is a pattern of government spending money on important projects and since a President said it was important, they had to scramble and come up with a category to spend the money on.

Not having a category can stop the government from allowing flesh colored hair on a driver's license, but it can't stop the government from spending money All this has culminated in the recent announcement of seven, earth-size planets around some ultracool star that could harbor water, and hence life, as we know it.

Some scientists see a pattern between planet size, distance from a sun, the possibility of water, and life. Of course, they assume that carbon-based life, like us, is what all living things in the universe would look like. Would humans even recognize a form of life that is not carbon based if we saw it? What about computer viruses? They reproduce and move about from place to place. Are they alive?

Personally, I have plenty to do trying to live life as presently constituted. As William Shakespeare said, "No, I will be the pattern of all patience; I will say nothing." I think I am rather good at saying nothing. Don't you? My wife wants to know in which context.

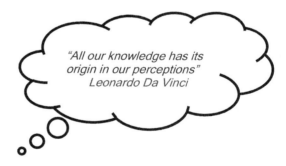

"All our knowledge has its origin in our perceptions"
Leonardo Da Vinci

PATTERNS: a poem

DIFFERENT

At some point you begin to realize that you are different.
Some people's lives are their own creation. At least they like to think so.
Mine isn't. I'm not entirely sure when I fully realized that I just didn't fit in.
Probably when I realized I was responsible for the sheep. I guess I'm slow.

It wasn't as if other things didn't matter once I got the sheep.
It's that the flock became the way I was going to accomplish the rest.
I got the farm so that the sheep would have a place to be born and grow.
So, the very purpose of the farm was to create the best.

Then it turned out that the farm was where I needed to raise my children.
So if the sheep didn't do well, then we would lose the farm.
Then there would be no place for the children to learn to care for the sheep.
Then there would be no sheep to save the family from harm.

It was my father's idea, or his fathers, I don't really know.
But he always said, "Nothing comes before the sheep!"
Because if the sheep aren't saved there is no farm or family.
It doesn't matter if you are tired, cold, sore, or lack sleep.

It wouldn't be any different if there had been some other way.
The purpose of the plan drives everything that follows.
The flock remains even as individual shepherds live and die.
Maybe the sheep care for the shepherds and teach them how.

PATTERNS: thoughts

PATTERNS: science

Do we create the patterns, or do they create us?

If something is random does it have a pattern?

If something has a pattern too complex to see is it random?

If we create something that is so complex that it appears chaotic to everyone else, but we know there is a pattern in its creation, is it chaos or not?

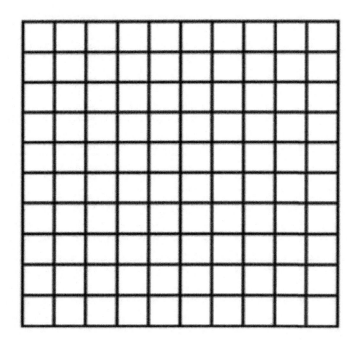

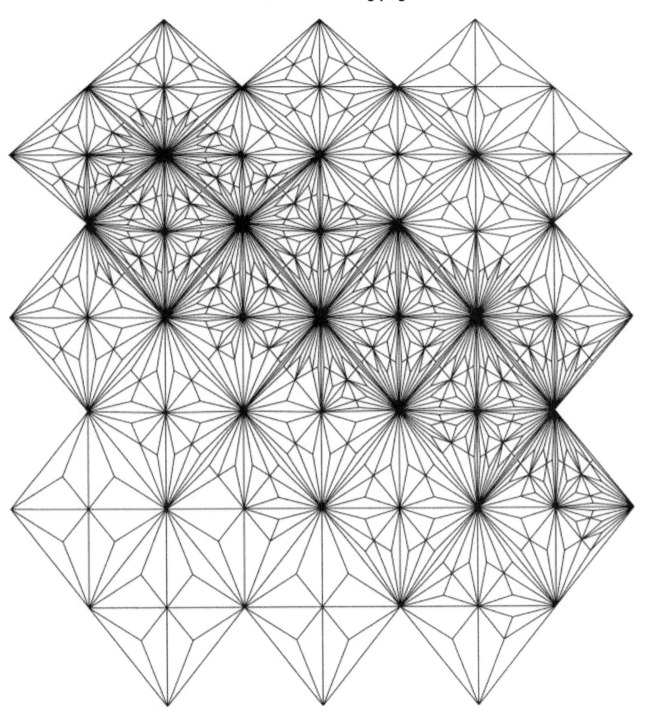

PATTERNS: sketch - art – draw - paint - paste – fold – etc.

Chapter 9

THINGS AND MINDS

*"Invention, it must be humbly admitted, does not consist
in creating out of void but out of chaos."*
Mary Shelly

At some point, "things" become "alive". There seems to be a continuum between non-living things and living things. No one knows, for sure, what life is exactly but, only in extreme circumstances, do we have trouble telling something that is alive from something that is not. Exactly when does a stack of material acquire that condition we call "alive"?

Most of us agree that electrons are not alive. Atoms do not appear to be alive. The interaction between electrons and atoms can make molecules like proteins. But most people don't think proteins are alive even though proteins make up cells which humans do think are alive. Then, some people don't think a fetus is alive; yet it is made of functioning, living cells. We don't always agree, but at some point, inanimate things start to be what we call "alive".

At some point, "things" also become a "mind". I don't know of anyone who thinks that an atom has a mind. Molecules do not seem to fit the concept of something with a mind. Viruses are especially confusing. They are only collections of molecules, but they can exploit a living cell and replicate. It is less clear whether a cell has a mind.

This confusion arises because in many ways' cells can do things you and I can do with our minds. Many cells move about in a purposeful way, detecting environmental conditions and responding to them. They seem to know what they like, and don't like, and how to meet their own needs.

A series of "things" can become more than just a collection of things if they are connected in a specific way. For example, the steering wheel on your car is connected in such a way that it can alter the direction in which you are moving. The steering wheel does not know what "direction" is, yet it can change your direction.

To do this, it must be connected to something else that it tells what to do. The steering wheel certainly doesn't know how it does what it does. It simply turns a shaft, that turns a gear, that pulls a rod, that shifts the axle. A collection of these "things" has become something besides what any of the things are separately.

Life and thoughts are kind of ghostly things. Life only occurs as lumps of stuff, and thoughts apparently only occur in brains - a lump of stuff within the living stuff. We express thoughts in words, but they are often just thoughts. When my wife asks me, "What are you thinking?" it is sometimes hard to tell her. A "penny for one of my thoughts" is probably never much of a bargain.

In the past, humans have treated "thinking", and being "alive", like a mysterious box. You can do something to the box, and then watch the result of what the box does in response. There is still no way of knowing what is going on in the box.

We know a lot more about the pieces inside the box now than we have in the past. But we still don't know what the critical point is when the box ceases to be alive, or to think. Sometimes we can even lose the ability to think while still being alive. I've been accused of that.

In the last century, Kurt Godel, Alan Turing, and others have tried building mechanical minds. This line of exploration has given us computers. Computers are less ghostly than minds because we understand the things with which they are made and how they fit together. But the things they are made of are organized in a very different way than our bodies and brains appear to be.

So, here I sit thinking, and I haven't a clue as to how I am doing it. I don't know if humans will ever be able to understand how things become "alive", or how things can make a "mind" that thinks. We still don't know exactly how, at some point, living things stop living and become just "things" again. When a mind stops thinking, what becomes of the thought?

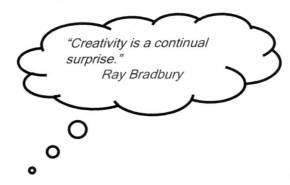

"Creativity is a continual surprise."
Ray Bradbury

THINGS AND MINDS: poem

CURRICULUM
Take the children from the cities
Let them care for living things
Especially the young
New in the blooming spring

Introduce them to the masters
Put tools into their hands
Teach them to grow the golden grain
And love the ever-changing land

Working together in the sun
Is the best way to find a friend
Let them face the storms of life
That never seem to end

Teach them how the arrow flies
To know there is wrong and right
Teach them to use the sword
So they never fear at night

Feed them with work and toil
For their dreams to lean upon
Show them love for beast and field
Before both are gone

Guide their hands across the wood
To feel the texture of the grain
Prepare the spindle and the wheel
Gather root and stem for stain

Guide them to the tailor's bench
Or perhaps the potter's wheel
Let them experience something real
And know how that can feel

Show them that great cities
Are not all that mankind needs
But men who love the beasts
And men who love the trees

Help them feel the peace and joy
Of caring for the land
The riches of the wilderness
Only come from the hand of man

THINGS AND MINDS: thoughts

THINGS AND MINDS: science

Do things come from the mind?

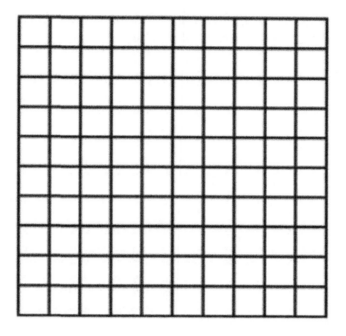

Do ideas in the mind come from things?

What is the difference between matter and material?

THINGS AND MINDS: coloring page

Humans invented swords, but not the concept of increased pressure when the point of pressure is reduced. That came from claws and teeth. Humans learned to fly, but not the laws of physics that enable flying.

But humans invented the concept of a flying sword as a symbol of something. What?

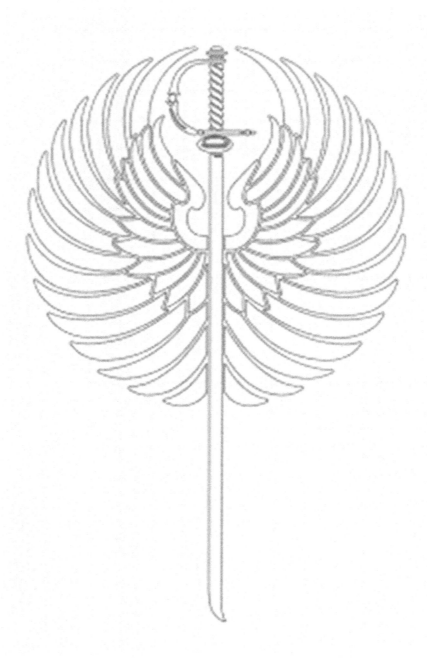

THINGS AND MINDS: sketch - art - draw - paint - paste - fold - etc.

Chapter 10

ANTICIPATION

"Thank goodness I was never sent to school;
it would have rubbed off some of the originality."
Beatrix Potter

You think you know someone well, only to discover new things after years of association. For example, my wife frequently reads to me while we drive. However, she's never been a big mystery fan, and I am. She relented recently and read a series of mysteries to me.

Things didn't go entirely smoothly. I discovered that we approach murder mysteries very differently. I was surprised because I thought we agreed on almost everything. (My wife disagrees with that statement, but this time I am putting my foot down and leaving it in.) Anyway, she kept trying to figure out the mystery in advance and made many guesses about what was going to happen next.

She'd say things like, "I know. It's the guy in Parliament isn't it?" To be honest, I had already read the books and knew the answers, so it was awkward. I knew she was wrong, which is a very dangerous position for a husband to be in. I finally discovered I could use a tactic that was being used by the attorney general in the news at that time. I would always reply, "I cannot comment on ongoing investigations." I thought it was quite clever. She didn't. I hadn't anticipated that.

Of course, she was equally surprised by me. She had assumed that, as a scientist, I would always be trying to figure things out. Murder mysteries are all about figuring people out, but science just can't do that.

I never try to guess the solution to a mystery. I just like to see where the story goes. I figure the whole idea of a murder mystery is to trick me, so why should I try? I just enjoy the story. Humans are so complex, interesting and unscientific. She anticipates. After fifty years of marriage, it was fascinating to learn that we approach things so differently in that way.

The difference between us probably explains why she is so well liked and socially successful and I am not. She actually thinks before she says anything. I have seen her spend several minutes thinking about a conversation before she makes the phone call. Sometimes she even tells me what to say when I am calling someone. The problem is there are so many possible ways to mess things up that I just get confused.

I don't really anticipate the effects of anything I say or do. What? You're surprised? I just say, or do it, and see what happens. I think it's the scientist in me. Simply do the experiment, collect the data, and analyze later. No chance for fudging the data that way. Research is full of surprises.

I guess anticipation can be fun too. Never having much experience with it, I can't say for sure. But it is deadly when applied to scientific experiments. While a scientist needs to be clear about the question being asked and the significance of the answer, it is very dangerous to anticipate the outcome or results. Such subtle bias has sometimes led scientists to fudge their data just a little.

That's alright, of course, if you are sure you are right like most scientists are. I mean, even Gregor Mendel of genetic fame apparently did a little of that sort of thing. But he turned out to be right, so it was okay.

Problems can also occur when someone pays you for your data and you know what kind of results they would like to see. You might feel you should fudge just enough to encourage them. It's a slippery slope though because you can't fudge enough to prove their point. Otherwise they won't hire you to do more experiments. Science is "a chancy job, and sometimes a little lonely." - Matt Dillon -

I guess anticipating what is going to happen is a valuable day-to-day skill, but a dangerous scientific attribute. This may be yet another way that scientists seem maladjusted.

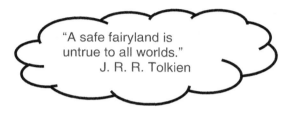

"A safe fairyland is untrue to all worlds."
J. R. R. Tolkien

ANTICIPATION: a poem

Ask the question
Design the experiment
Finding clean glassware
Is quite an accomplishment

Clear off a lab bench
Assemble apparatus
Order needed chemicals
Schedule time continuous

Initiate trial
Time meticulously
Using pre-prepared logs
Record data scrupulously

Plot raw data
Carefully calibrate
But whatever you do
Don't anticipate!
SCIENCE

ANTICIPATION: thoughts

ANTICIPATION: science

In what ways do science experiments and art projects seem alike?

How can a graph aid in both art and science?

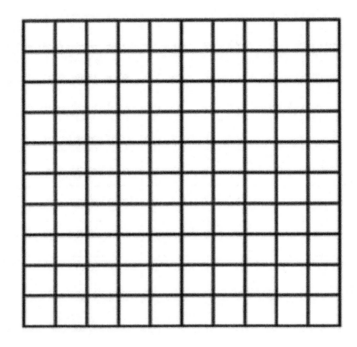

Does an artist know what his final product will look like before he begins?

Does anticipating a certain end-product in art inhibit the final product, or enhance it?

ANTICIPATION - coloring page

Much of art developed trying to control the physical world to make something useful, beautiful, or both. Much of science developed trying to predict or control the physical world. It turns out the two are useful to one another.

However, prediction is a very tricky business. The weather service has made great strides in understanding patterns and using that understanding to make predictions, which are frequently inaccurate. Yet they can be pretty, and sometimes useful.

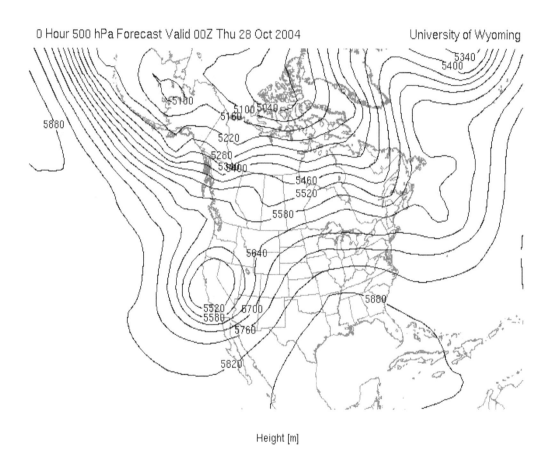

Height [m]

ANTICIPATION - sketch - art - draw - paint - paste - fold - etc.

Chapter 11

THE IMPORTANCE OF TRUTH

"The Light failed; but the Darkness that followed
was more than loss of light."
J.R.R. Tolkien

Something can be both true and unimportant. Not everyone seems to be able to make that distinction. Of course, the meaning of importance may be as hard to come by as the meaning of the word true. It all depends on what the definition of "is" is.

Now that I'm thinking about it, I guess there can be things that are false that are important. Politics for example. Oh, and there can be things that are false that are unimportant. Like, . . . politics. Oh, and something can be true and important. For example, my wife is still with me. Things get confusing quickly, at least in my mind.

I once spent the better part of two years of my life determining that a specific drug interfered with the embryonic development of a parasitic worm. I suppose that was very important to the worm, but it hasn't proven particularly important to humanity. I like to think of it as a should that someone will one day stand upon to make a great discovery.

I don't feel too bad about the time and effort though because it was a learning experience which was important to me. I was able to utilize the expertise I achieved in research on that project to later spend another couple of years determining that a specific insect parasite didn't really cause all that much harm to the insect. Really, that's true. Both that I wasted the time doing that and that the parasite really doesn't harm the insect much.

Some chemical dyes are photosensitive. When they are illuminated, they give off an electron and cause singlet oxygen to be released. Singlet oxygen is highly reactive with other compounds and can cause a lot of tissue damage. This phenomenon has been used to treat some cancers. They flood the tumor with dye and then shine a light on it which kills the cancer cells.

Anyway, a student and I were able to determine that the dye Rose Bengal, in water, kills mosquito larvae when they are then exposed to light. Unfortunately, when Rose Bengal in water is exposed to light the singlet oxygen can damage or kill most everything else in the water also. This makes it an unimportant method of mosquito control, even though it is true that it will damage mosquito larvae.

I could go on . . . but it would be embarrassing. It isn't that I have had any problem with determining what is true or not. It's just that I seem to have a problem determining what is important to do research on.

I also seem to have trouble determining what is important to do daily. It just isn't clear to me whether I should finish refinishing the cabinets, work on surveying insect parasitic nematodes, build another mountain dulcimer, or play the guitar. My wife has tried to help me develop my prioritization skills, but she is getting frustrated.

She has suggested that to set proper priorities I need to be more in touch with reality. That doesn't make any sense because I am a scientist. Can a person be any more in touch with reality than scientists who study reality? I mean those worms and mosquitoes were really dead.

This is the age of science! We are flooded with facts and reality. Every problem has a scientific solution. At least everyone always claims that science supports their solution, just like everyone claims their problem is important.

Determining what is true is easier than determining what is important. It may take a lot of time and effort, but one can usually come to some understanding about what is true, even in politics. Some people think that everything is relative, but if everything is relative then that statement is relative. If that statement is relative, then something is true.

However, what is important seems to be a moving target. Under one set of conditions just staying alive may be the most important thing you can apply yourself to. Under a different set of conditions risking death may seem worth the risk. In some situations, playing the guitar may be the most important thing you can do. o

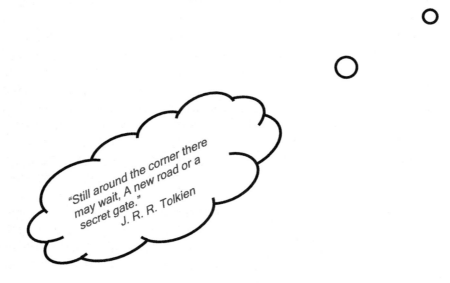

"Still around the corner there may wait, A new road or a secret gate." J. R. R. Tolkien

THE IMPORTANCE OF TRUTH: a poem

THE IMPORTANCE OF TRUTH

Truth and Importance went for a walk
Decided it was time they had a little talk
No one said much for quite a while
At least the better part of a quiet mile

Finally truth spoke up and said with a grin
Importance is important but it's not my twin
I appreciate that we've long been friends
But it's clear Truth's more important in the end

Importance thought carefully for quite a while
Importance is important and that was his style
Finding the truth, he said, is all very fine
But I wonder which one takes more time

I admit that it's hard to tell Truth from a lie
But what's more important after we die?
The truth about what he and she said
Won't matter too much after we're dead

THE IMPORTANCE OF TRUTH: thoughts

THE IMPORTANCE OF TRUTH: science

The essay talks about the difference between what is true and what is important? Does this apply to art?

How could an artist portray this truth? Importance? A lie?

How many examples can you list of this dichotomy in either field?

Which is more lacking in everyday experience; truth or importance?

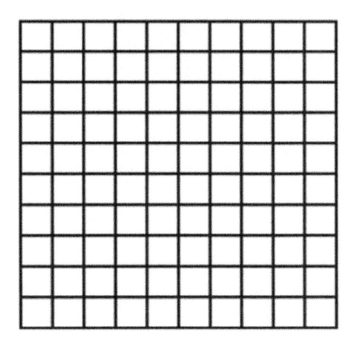

THE IMPORTANCE OF TRUTH: coloring page

Fibonacci numbers a series of numbers in which each number (Fibonacci number) is the sum of the two preceding numbers. The simplest is the series 1, 1, 2, 3, 5, 8, etc. The following illustration is made in the Fibonacci sequence. This sequence is found repeatedly throughout nature: helix shells, sunflower seeds in the flower, pinecones, etc. All that is true so it must be important. It is not clear why it is important.

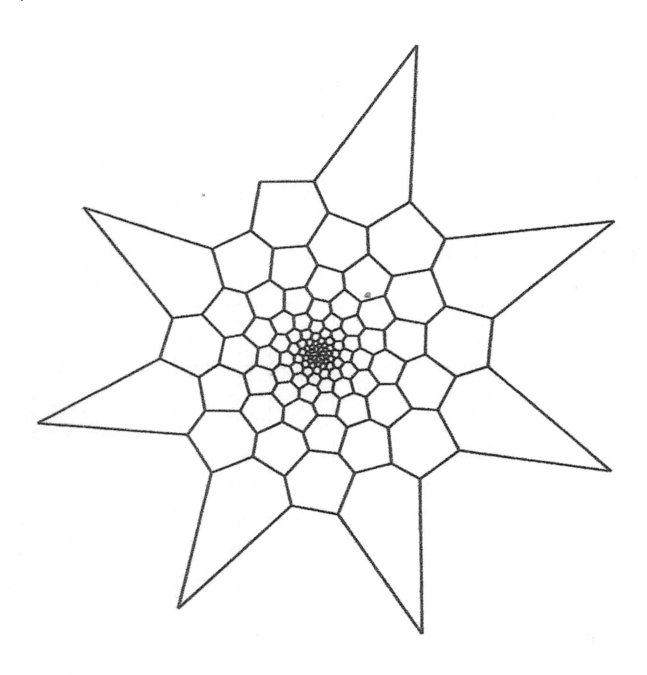

THE IMPORTANCE OF TRUTH: sketch - art - draw - paint - paste – fold – etc.

Chapter 12

THE SHAPE OF THINGS

"If a person has a contribution to make,
he must make it in public.
If learning is not made public,
it is a waste."
Chaim Potok

I am trying to get into shape, and the shape I've chosen is a triangle. (The old jokes are best, aren't they?) I am still longer in one dimension than the other, but the dimensions have kind of reversed. Anyway, I have been successful in changing shape, even if it isn't pretty. More recently, I have been trying to size things up.

Physical matter always has shape . . . and size. When the first alchemists started thinking that there were some kinds of invisible particles that made up matter, it followed that those particles had shapes and sizes.

At first, it was difficult to discern the shapes of invisible things. Actually, it's still difficult to discern the shapes of invisible things. But if invisible things are made of material, they must have shapes; and if shapes, size.

It stands to reason that, if you don't have a size, you aren't in any kind of shape. Having a size is the first step of being in the shape you are in. We should be grateful for the shapes we are in; at least, as far as that goes.

It's a lot easier to determine the size of invisible particles than their shapes. If we put "stuff" on a plate of Jell-O, particles of the stuff will diffuse out from where there is a lot of it to where there is less. You can watch it spread and measure how far it has moved over time. It's boring, but you can do it. Scientists do boring stuff like that all the time. Next week I'll tell you about the excitement found in sorting crystals.

Back to size, small particles seem to move faster than large ones, but it's because they can more easily fit in between the particles of the Jell-O. It's like when my young sons were able to move out of the chapel, after church services, faster than me because they were little and could dart between people. I'd have been arrested for trying the same moves. The people who thought I should control my sons better just didn't understand physics.

Since little particles diffuse faster than large ones, we can calculate relative sizes of invisible particles. It was extremely tedious to test all invisible particles against each other, and it took a peculiar kind of person to do something like that. Now days we call them scientists but, back then, they were just peculiar.

The dirty little truth is that science is a lot of work, and it's usually tedious. Everyone concentrates on the "aha moments" and ignores the collection of 780 data points collected over two years to reach an insignificant conclusion. That nicely describes my master's thesis research.

The only thing more tedious than collecting my master's data was retyping the results twelve times to get the manuscript right. My wife did that for me. I think she thought I was going to make a lot of money from that degree. Oh well, she lived and learned. In reality, no one would watch a movie about how science is really done.

Determining size is only the first step. The entire point of chemistry, if chemistry has any point, is that invisible things interact with one another. For material things to connect, they must get close together; and their closeness depends not only on size but also shape. So, I guess some chemicals must have a point. Other chemicals may be more blunt.

I took my first chemistry class in the 1960's. I thought all the attention being paid to invisible things was a communist plot to distract students from important stuff like the Viet Nam War and Civil Rights. As I examine the shape the world's in now, chemistry seems a lot less complex. Today it seems like politics may be a communist plot to keep people from thinking about important things like chemistry.

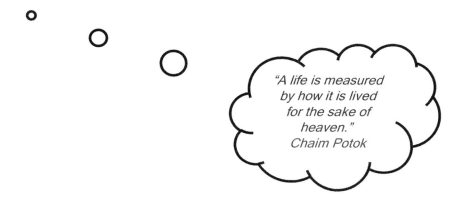

THE SHAPE OF THINGS: a poem

There are dusty shapes floating in the sun
But when I try to see them, they come undone
I guess that's why I'm restless trying to find
The things I see floating in my mind

The truth of dust pours out the door
And I almost hear footsteps across the floor
As if I haven't chased dust in the past
Only to find she moves too fast

From dust to dust for dust thou art
The dust fills and chokes my heart
And as the dust disappears
I can no longer see you here

So what are these shapes floating in the dust
And why do I feel compelled to do them just
Dreams and visions can't be true
But can they be important too

THE SHAPE OF THINGS: thoughts

THE SHAPE OF THINGS: science

What is the shape of truth?

What is the color of importance?

How does true sound?

Is there a direct correlation between true and important? Can you graph it?

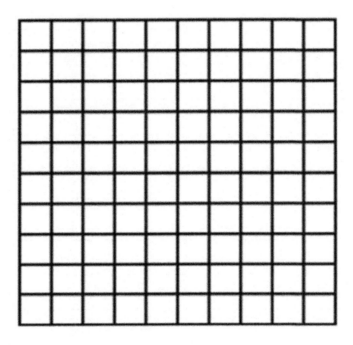

THE SHAPE OF THINGS: sketch - art - draw - paint - paste - fold - etc.

Chapter 13

THE IMPORTANCE OF PEOPLE

*"The road to creativity passes so close to the madhouse
and often detours or ends there."*
Ernest Becker

They say that old men become obsessed with the past. I don't know who "they" are, but my wife is one of them. She's wrong though because I'm not. I don't even know the geologic time scales. I certainly have done my best to forget about the past, and I deny the past when she brings it up.
I don't believe that time is as significant as people.

For example, "they" also say that man has been on the earth for one-hundred thousand years. But if that is true, I would sure like to know what man was doing for the first ninety-four thousand years. They apparently weren't developing language, math, poetry, music, culture, or art.

If we go back, beyond about fifteen thousand years, there are almost no archeological markers at all. That means, for the first eighty-five thousand years, people built nothing and left nothing behind. They apparently lived a subsistence-level existence and left no artifacts. They may have experienced our same emotions or views, but we will never know.

Then suddenly, about fifteen-thousand years ago, people got their act together and started making wheels and tools. They began a form of agriculture, though there was still no written languages or math. Then something interesting happened again about five or six thousand years ago. Language and math seemed to leap into existence all around the world within a relatively short period of time. Well, six thousand years seems long to me, but relative to a hundred thousand, it was the blink of an eye.

Who were the intelligentsia that started us down the road to unprecedented wealth, health and well-being? Don't look at me. I didn't do it! Some might even say I've been an impediment.

Apparently, intelligence and culture aren't a function of time, but of people. The Population Reference Bureau estimates that the number of people ever born is one hundred and five billion. But only about two percent of all people were born before the birth of Christ about two thousand years ago. So only about two-billion people lived in the first ninety-four thousand years.

That means that ninety-eight percent of everyone who has ever lived has been born since then. After a ninety-four-thousand-year slow start, humans have created everything from the pyramids to Proust, Newton to nanotechnology, salt mines to Shakespeare, and cave art to cubism. Okay, I guess cubism isn't all that great; but still, it's a pretty good record.

Consequently, it seems to me that we don't need more time to solve our problems. We need more people. All the swell ideas of the last few thousand years came from people. I know there are those who say we have too many people, but the people who say this seem to be the same ones saying we have too many problems. The solution to past problems appears to have been people.

It's even taken people to devise the means for reducing the number of people. Atomic bombs, abortion, and birth control pills were all invented by people. Wouldn't it be ironic if humanity invents the means to do away with humanity? Tragic might be a better word since the world before intelligence, I believe, was a brutish place.

I'm told there might be a global, environmental catastrophe developing. I guess getting rid of people could avoid that. Perhaps going back to a brutish existence might help. I can't think, however, that increasing the number of intelligent people might be a better solution . . .

Personally, I don't think old men are obsessed with the past. I think they are obsessed with the future. Maybe, more accurately, the lack of their personal future. Considering the current state of affairs, I wonder what the world will look like next year. Can you imagine how it might look in another ninety-four thousand years? If there are no people, I suspect it will be brutish.

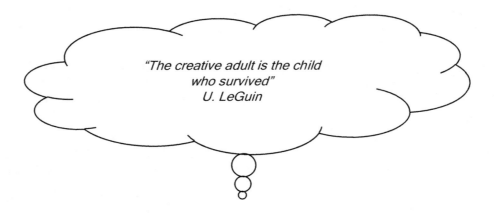

"The creative adult is the child
who survived"
U. LeGuin

THE IMPORTANCE OF PEOPLE: poem

NO PLACE LEFT TO GO

Everybody just wants to have their way made clear
Should we leave or maybe just stay around here
Where can we go to find the warmth of the sun
Just waiting to hear from the anointed one
Like Prophets of old
We'd all like to know
We'd all like to be told
But the truth is there's really no place left to go

In days of old things just weren't the same
Prophets of God spoke, you all know their names
There was someone to tell us, give us fair warning
Lead us into the wilderness first thing in the morning
Like Prophets of old
We'd all like to know
We'd all like to be told
But the truth is there's really no place left to go

Everybody's going somewhere just as fast as they can drive
I guess they're just trying to do what they must do to stay alive
You know I'm just like that Baby running away from me
Always just a day away from the place where I want to be.
I need a Prophet of old
I'd just like to know
I'd like to be told
But the truth is there's really no place left to go

THE IMPORTANCE OF PEOPLE: thoughts

_____

_____

_____

_____

_____

_____

_____

_____

_____

_____

_____

_____

_____

_____

_____

_____

_____

_____

THE IMPORTANCE OF PEOPLE: science

Ideas only come from people. If animals have ideas, they have no way to transmit them. Progress of humans only come because of other humans. The more people the more ideas. The more ideas the more progress. So why abortion?

What would a graph of ideas versus population look like? Could such a graph be extrapolated for rising populations numbers? Is there a limit to population? Is there is a limit to ideas?

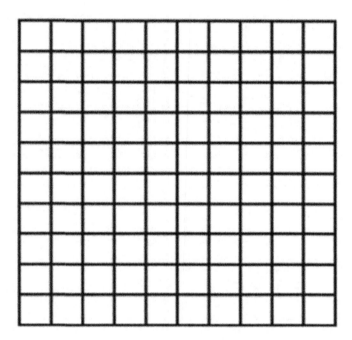

THE IMPORTANCE OF PEOPLE: coloring page

Humans make discoveries and think of new things. Even places and things that are not yet known. Every parent first imagines a new family.

THE IMPORTANCE OF PEOPLE: sketch - art - draw - paint - paste - fold - etc.

Chapter 14

STEM

"It is through science that we prove,
but through intuition that we discover."
Henri Poincare

My wife and I have an arrangement. I get big ideas, and she tells me why they won't work. It saves me a lot of time and energy because she usually helps me refine my idea into something smaller that can work. And she helps me stay with an idea until it's accomplished instead of jumping to another one. Well . . . no one is successful all the time . . .

Recently I got an idea about how to change the entire STEM educational system of the United States into a system that would better benefit mankind. By the way, STEM stands for science, technology, engineering and mathematics.

Once again, my wife pointed out that I have tried this before and it didn't work. So, reluctantly, I agreed with her and decided instead to focus on just writing a highly informative, for once, science column. She pointed out that I had tried that before also and it hadn't worked either.

Anyway, the problem with STEM education is that science and technology are two different activities with different purposes. Science discovers how the world works, and technology tries to control how the world works. There is a similar difference between math and engineering. Math discovers principles of logic, and engineering tries to use those principles to control the world.

There is a black spiritual, based on events described in the Old Testament that goes: "Ezekiel saw the wheel, way up in the middle of the air." Actually, Ezekiel saw two wheels, one inside the other (Ezekiel 1:16). When two wheels are set inside each other, they are made to turn together, in the same direction. The outer rim and the axle of a wheel is an example of a wheel within a wheel. Ezekiel's wheels were turned in the same direction and with similar purpose.

In contrast, William Blake wrote a poem entitled "And Did Those Feet in Ancient Time". This is the second stanza.

> *"And did the Countenance Divine,*
> *Shine forth upon our clouded hills?*
> *And was Jerusalem builded here,*
> *Among these dark Satanic Mills?"*

It is generally assumed, because of the time and place in which Blake lived, that he spoke of a wheel outside of a second wheel, as was found in an industrial mill of his day. Wheels in mills operate outside one another and turn in opposite directions. One wheel turns the other by intermeshing cogs. These two wheels are divided in space, direction of rotation and purpose. Apparently, Mills thought such an arrangement Satanic. It is, at least, the opposite of wheels within wheels.

These images illustrate, in a non-scientific way, the difference between science and technology, math and engineering. Science seeks to understand life and experience. Technology seeks to control life and experience. They do not work together.

The word "control" literally means "to roll against". When used as a mechanical principle, opposing wheels are excellent forms of control. I, myself, use them often. But as a metaphor for our culture, it suggests that sometime after Ezekiel, man began to see himself - not as a force working with creation, nature, and life - but as a force working against natural forces. Control is gained through engineering and technology.

Instead of combining these subjects into STEM, we should be separating them. The study of science and math are more aligned with the liberal arts which attempt to understand and explain the human condition. They are creative endeavors of logic and reason seeking understanding. In fact, the study of math and science are more in keeping with the arts since they both seek to understand and represent the world as it exists in a different way.

Technology and engineering are about materials, machines, and their practical applications. These subjects assume that because man can control the world and life that he ought to. They seldom address whether we should be controlling things. As such, they belong more in the social sciences or at least the school of business. Most technology ends up in businesses already anyway.

My wife thinks this is a brilliant idea with about as much chance of happening as my previous suggestions about science education. She suggests I refine my idea into how to finish refinishing the kitchen cabinets instead. I told her I was a scientist and that was more of a technological function. But I've tried that argument before, and it didn't work.

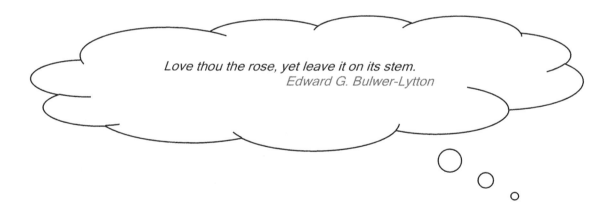

Love thou the rose, yet leave it on its stem.
Edward G. Bulwer-Lytton

STEM: a poem

THE BUSINESS PLAN

To stand at the chiasmus of two eternities:
Past and future. Arriving at the boundaries
In just the nick of time.
Notching and noting it on my walking stick,
Having made good time with little traffic,
And it being a gentle climb.

I hope you will pardon some detail obscurity.
There are more secrets in mine than in some men's story.
Not due to any crimes,
Just inseparable from its very nature,
The warp and the woof of its every queer feature,
Peculiar, odd pastimes.

How can I explain the search for a missing hound?
Inquiry of travelers, "Have you seen . . . nose to ground?"
I am still hot on the trail.
Or the assistance rendered to the bright, rising moon,
Though never to influence the time; late or soon.
Still, my presence did avail.

Self -appointed inspector of sun, rain and wind,
Patrolling sidewalks without complaint or chagrin.
Did my duty faithfully.
For a very long time, I may say without boasting,
I did what I did with without pay, never coasting.
Duty with humility.

But at last it became increasingly evident
That I lacked something, in the communities' judgment.
No room in the inn, it seems.
I would not be admitted as a town officer,
Nor given allowance for a place sinecure.
Self-responsibility.

So I have calculated a small business plan,
Counted, estimated the cost before I began.
Experience is helpful.
To manufacture dreams from simple sun, rain and wind.
And market them boldly across all continents.
Bright shiny beads and baubles.

STEM: thoughts

STEM: science

When art becomes a business does it become technology?

What differentiates art from technology?

Is it an artistic endeavor when science discovers a fundamental law of the universe?

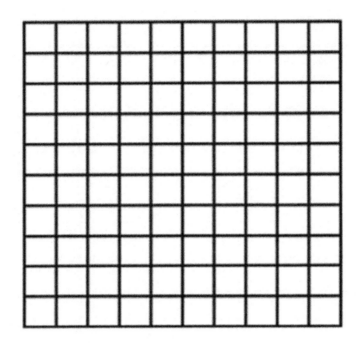

STEM - sketch - art - draw - paint - paste - fold – etc.

What can I say? Flowers have Stems.

STEM: art - draw - paint - paste - fold - etc.

CHAPTER 15

OPPOSITION IN ALL THINGS, too

"For if joyful is the fountain that rises in the sun,
its springs are in the wells of sorrow
unfathomable at the foundations of the Earth."
J.R.R. Tolkien

The shortest day of the year will be December 21. The solstice will occur around 3:23 PM mountain time. This is the exact opposite of the summer solstice which will occur on June 24, at approximately 9:45 AM. Opposites are important in science as they are to everything. If you have any two things that are distinctly different, you can turn that difference into language that can express anything you want.

I'm not sure when someone first discovered the power of opposites. But I think Samuel Morse was the first, I know of, who put opposites to use when he invented Morse Code; the series of dots and dashes used to send messages by telegraph. You could write the complete works of Shakespeare in Morse code and anyone, who can read the code, could read the book. It might be a little bulky.

I am writing this column in a code made of ones and zeros with the computer. Whenever something doesn't make sense in one of my columns, as frequently happens, I blame the code. My wife proofreads my columns and finds lots of mistakes where the code was wrong.

There are lots of opposites used in the scientific world besides the winter and summer solstice. Many chemicals exist in opposite forms, which could make them excellent languages, if anyone could understand chemistry. We didn't know about chemical opposites until only a little over 150 years ago when Louis Pasteur discovered "isomerism".

Louis Pasteur was a French scientist mostly known for his work in microbiology. In fact, the French call hm the father of microbiology. The Germans, of course, disagree saying that title goes to Robert Koch. But Pasteur trained as a chemist and made one of his most important discoveries in that field.

Early in Pasteur's career he studied the chemical, optical, and crystallographic qualities of a group of compounds called tartrates. Tartrates aren't just another boring compound. Especially when you realize that tartrates are critical to wine production and Pasteur was French . . .

Anyway, natural-occurring tartrates, like those found in wine, rotate the plane of polarized light as it passes through the crystalline form. However, tartrates created synthetically don't rotate the light in the same direction even though their chemical formulas are identical as the natural tartrates.

Pasteur, while examining crystals under a microscope, noticed that some tartrate crystals appear to have a geometric property known as "chirality". Some crystals appeared to be "right-handed" and others appear "left-handed". That means the faces of the crystals cannot be superimposed on their mirror images. Further examination proved that it was this property that determined how the two forms of the molecule twisted polarized light differently.

This opposite geometric orientation, of some molecules, is important to how we smell and taste; not just wine but other foods as well. Another example of" optical isomerism" is a compound called carvone which occurs in nature in two isometric forms: left-handed carvone and right-handed carvone. The left-handed version is found in the dill herb, and the right-handed form is found in spearmint. My wife likes spearmint gum and dill pickles, but I don't think she would like dill gum.

The mechanisms of smell and taste are not entirely understood. Presumably, however, chemical receptors work a little like locks and keys with some sensory receptors accommodating a left-hand molecule and others a right-hand molecule. There may be no receptors for certain chemicals in some individuals. It isn't known if there are ever ambidextrous receptors.

The entire universe seems to be a collection of opposites such as winter and summer, forward and backward, up and down, right and left, and dill and spearmint. Opposition in all things seems to be a universal principle of science as well as the basis for all blues music. I wonder if the winter and summer solstices form a cosmic message to, or from, the universe. We just haven't translated it yet.

"You see, the point is that the strongest man in the world
is he who stands most alone."
Henrik Ibsen,

OPPOSITION IN ALL THINGS: a poem

ETERANL OPPOSITES

We live in the present tense
Sweet children of innocence
But we also lived once before
Though the past seems ancient lore
And then time will take its toll
And we'll live again as eternal souls

What has become of all the years?
What will become of future tears?
We sit and wish upon a star
Just who do we think we are
Does the dark turn into light?
Does the light end with the night?
So if death is really a thing
There must also be a reawakening

How is the concept of opposition expressed in science and technology?

How many "opposites" can be found in the graph?

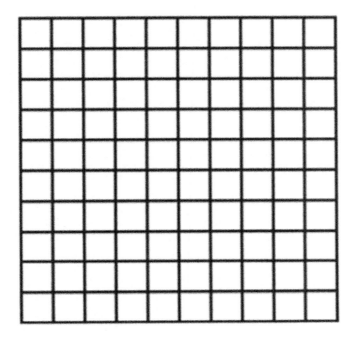

How is the concept of opposition expressed in art?

How many degrees of opposition are there?

--- .--. .--. --- ... .. - .. --- -. .. -. .- .-.. .-.. - .... .. -. --. ...

(in Morse code)

Over and under, up and down, forward and back, right and left, black and white, zero and one, crooked and straight, odd and even, ????

OPPOSITION IN ALL THINGS: art – draw – paint – paste – fold – etc.

Chapter 16

WHATS THE MATTER WITH MATTER?

"Energy is liberated matter,
matter is energy waiting to happen."
Bill Bryson

I think there is something the matter with matter. While matter itself seems to matter, the questions about matter seem to matter more. Is the material world infinite? If not, what is there past it? Do places without matter, matter? Is matter infinitely divisible into smaller pieces of matter? My wife thinks it doesn't matter.

Rabid materialism is part of modern life and is, to a great extent, the result of science. It's just a fact that science is materialistic because it must be. The purpose of science is to help us negotiate the material world.

It is a little odd that science claims to use "reason", which is not matter at all, to reach accurate conclusions about material. You might say that reason is immaterial. I guess that is why many scientists are trying to discover a material cause for reason by looking for it in the material of the brain.

Don't be too disappointed in scientists about all this confusion. We're only human, so it stands to reason that we would be a little confused sometimes about what matters. My wife thinks I'm particularly good at being confused about what matters, for a scientist.

The brain is made up of matter and yet it produces ghostly things like "reason". How it does this is a mystery. The brain itself appears to be mostly for negotiating the material world because four of its five lobes are devoted to sensing and responding to sensory information about the material world.

The frontal lobe seems to be the one where we "think", although thinking may not be what we think it is. All those things that we think about, but can't be held in our hands, like reason, are abstractions. The frontal lobe appears to be where we make abstractions like "reason," which aren't matter. How does it do that?

And whenever we try to discuss abstractions, or use them to explain ideas like reason, we discover that we frequently must use materialistic words to describe immaterial things. That is why we talk about loud colors, backward thinking, high aspirations, low brow people, and political right and left. We talk about rich possibilities and poor reasoning. One would think there would be a "reasonable explanation," which is using an abstraction to modify an abstraction.

No one exactly understands where ideas like past, future, memory and planning originate, since we have no sensory system for identifying time outside of our normal senses. Our bodies do respond to time in a cyclical manner, but our brain doesn't control that. The ideas of past and future are non-material. My wife likes to point out to me that "Getting around to it" is not a real thing either.

The brain does so many things besides sensing and responding to the immediate environment that we assume there must be something else in there somewhere. We call that "something else" the mind. That's what any good, materialistic person would say. We give abstractions names as if they were matter, actual objects.

Anyway, I've come to realize that I am a product of our modern educational system. I am highly educated about abstractions which aren't made of material, but I can't really "do" anything materially.

Well, I can do some things. I can dissect tiny insects with precision. I can wield a mean microscope using multiple means of illumination. I can identify a rather impressive number of creeping and crawling things. And after years of teaching anatomy and physiology I am one of the few people who probably knows all the ways there are to skin a cat. These are not skills in high demand.

The brain has no known way of sensing immaterial things, let alone knowing how to respond to them. That's what seems to be the matter with matter. While the material world seems to be all that matters, the immaterial world of reason and abstraction is totally "non-sensible."

"All have their worth and each contributes to the worth of the others."
J.R.R. Tolkien

WHATS THE MATTER WITH MATTER: a poem

NOTHING MORE

They say it's yours
Words and nothing more
This will last forever
Quoth the Raven, Nevermore
Solid as a rock
As if rocks were sure
Seeing is believing
Just so much manure

Matter isn't matter
Energy and nothing more
Matter doesn't matter
It's just a metaphor
The things that really matter
The things we all die for
Is seldom really matter
It's love and nothing more

WHATS THE MATTER WITH MATTER: science

What is the difference between matter and material?

How can one measure the difference between matter and material scientifically?

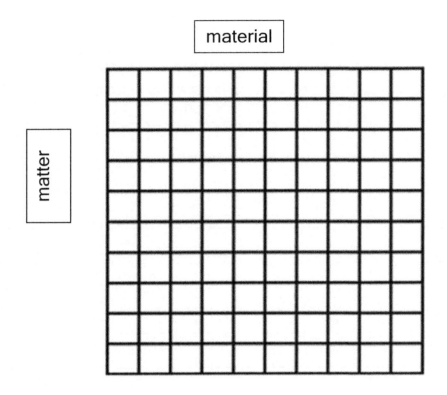

Does art use matter to create material? How could one show the difference between material and matter in art?

WHATS THE MATTER WITH MATTER: coloring page

I call this the spiders harp. All things are made of matter. Living things are made of matter. However, music and art are also made of matter. This seems a bold interpretation of matter.

WHATS THE MATTER WITH MATTER: art - draw - paint - paste - fold - etc.

Chapter 17

F-HOLES

I wonder if a soldier ever does mend a bullet hole in his coat.
Clara Barton

I have been thinking about the f-hole on a violin. I mentioned this to my wife, and she asked, "Why?" She asks the oddest questions sometimes. I explained, of course, that I wondered why guitars usually have round sound holes. Some jazz guitars have f-holes too, but most electric guitars don't have sound holes at all. Why wouldn't someone think about that?!

I always thought the sound hole was there to let the sound out. I guess that's true, except that sound is just a wave of air molecules. So, what the sound hole lets out is air. The air is pushed out by the vibrating wood which vibrates because of the vibrating string.

However, if the wood is vibrating, that means that the wood is pushing one way and then another. When the wood compresses the air inside the sound box of an instrument, it pushes air out. When it springs back to its original position, it sucks air back in. This is what creates the oscillating waves of air we call "sound".

The sound box of an instrument holds more air than can easily pass through the smaller sound hole. If the sound hole is round, most of the air exits and returns through the middle of the hole. However, the air along the edge of the sound hole is trapped between the air compressed inside the sound box and the mass of air oscillating through the center of the hole.

This same thing happens around the wing of an airplane. The sound rushing over the top of the wing is compressed by the air above it. It's like putting your thumb over the garden hose. The water that then must push through the smaller diameter has to move faster and with greater force.

In the case of a stringed musical instrument, the air exiting around the edges of the sound hole must move faster than the air through the center because it is compressed by the mass of air in the center and the solid edge of the sound hole. This means that the air around the edges of the sound hole must account for a greater portion of the sound wave than the air through the center of the sound hole.

Does this account for the surprising volume of a violin as compared to an acoustic guitar? The f-hole provides more surface area for the Helmholtz effect to work than a round hole thereby increasing volume, especially of lower tones.

The f-hole was practically born with the violin in the 1500's. A few of the earliest violins had C-holes, but the design by the Amati family was very quickly adapted and evolved to the f-hole. Helmholtz didn't make his discoveries until the late 1800's, so what drove the early adoption of the advantageous f-hole?

Today's emphasis on progressivism often leads people to believe that previous generations were somehow "less enlightened". However, the people in Cremona, Italy, where the early violins were created, did not know about Helmholtz. But Cremona was a multilingual, sophisticated society that was steeped in the mathematics of the Greeks with its emphasis on music, geometry, and proportion as studied in the Quadrivium.

They understood that a sphere could be "peeled" into a spiral, and the resulting peel would be, in form, a cursive "f". If the sphere was a hole, the shape of a peeled hole would be an f-hole. While they might not have understood the Helmholtz effect, they understood surface area, form and function perhaps to a higher degree than the average person does today.

Progress implies moving towards a goal, in the case of the violin f- hole the goal was greater volume and better tone. It isn't progress if the goal keeps changing. Now I wonder why more stringed instruments don't have f-holes, whether we have progressed as much as we think, and why I didn't I think about this sooner?

F-HOLES: a poem

F - Hole

The mark of Seth is carved in me
A cold wind blows down my back
I shiver in sweet agony
The figure all dressed in black

I turn and look all around
All eyes look right through me
Tears at the sweetest sound
With their ears they see

I see what can't be seen
A voice that speaks no words
I am the wind that runs through me
And the longing afterwards

F-HOLES: science

Why is the scroll style "F" used as a sound hole for all violin type instruments?

The f-hole was born from art and only verified by science later.

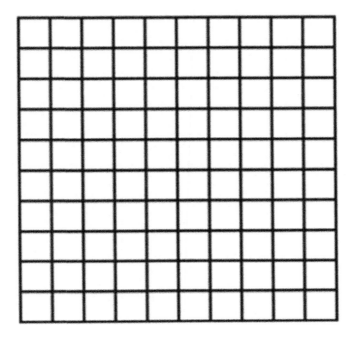

Are there things verified by science that were later utilized in art?

The discovery of zero was a significant event in History. How do artists use the concept of zero?

F-HOLES: art - draw - paint - paste - fold - etc.

Chapter 18

BEAUTY AND BEASTS

"The question is not if we will be extremists,
but what kind of extremists we will be.
The nation and the world are in dire need
of creative extremists."
Martin Luther King Jr.

It's my wife's fault for having too many grandchildren. How was I to know that having children would end up being so complicated? I just wanted to have fun playing music with my grandkids, so I started building mountain dulcimers for them. Dulcimers are simple for children and adults to learn to play.

But the grandkids keep coming and it's all starting to get expensive. Do you know how much hard woods cost? Then I discovered a beautiful wood right here in Colorado that is less expensive and can be used to make dulcimers: blue-stain pine!

Nature often creates beauty through natural processes that seem ugly or harsh. Take a brightly colored mushroom growing amidst decaying organic matter, for example. Or a gold, metallic-colored beetle rolling a ball of dung. The geologic terrain is shaped through violent storms, earthquakes and volcanoes. In the natural world, one lives when another dies. Vegetation is eaten to sustain life, and waste is produced to sustain vegetation.

In recent years, Colorado has experienced an infestation of *Dendroctonus ponderosae*, the mountain-pine beetle. Numerous trees in our pine forests have succumbed to this infestation. Their grey shapes can often be seen looking like ghosts in the dense green foliage of the forests.

The adult, female beetles chew through the bark of available trees while releasing pheromones, a gaseous attractant for male beetles. They mate and lay eggs within tunnels chewed into the trees. The eggs hatch, and the larvae continue to eat and burrow through the trees. Eventually the larvae pupate and hatch out the next year as new adults.

Once trees are infected, pheromones attract more beetles to the same trees resulting in a mass infestation that can eventually kill them. One cause of trees death is the numerous galleries and tunnels through the trees that inhibit water and nutrient flow.

Trees have some natural abilities to combat such infestations. One defense mechanism is the resin that's secreted by pine trees, commonly at the site of wounds. However, pine beetles carry, within a pouch in their mouthparts, a variety of Ascomycete fungi that can interfere with the tree's resin production.

The fungus produces a thread-like mass of "hyphae" and spores that are "sticky". These eventually block the water-conducting columns of the tree, drain the trees of their nutrients, and eventually cause the tree to starve to death. Beetle larvae feed on the fungus, as well as the tree, further promoting beetle growth and increasing their numbers.

The fungal hyphae are made of inert material that persists after the fungus dies. These colored hyphae cause discoloring of the wood. While the wood is often discolored bluish, different fungi can stain the wood other colors, including <u>shades</u> of blue, brown, green, red, and even black.

If a tree killed by pine beetles can be harvested quickly though, the wood can still be used. However, with too much time lapse the wood begins to dry, crack, and warp due to the many tunnels left by the beetle. These problems make it difficult to use the lumber for building. Many of the infected trees infected are older, with diameters greater than eight inches. They are often scattered within the forest making it difficult to harvest them.

The death of an ancient tree is sad, but natural. The association of three living organisms that complete their life cycles in tandem is an arrangement called symbiosis, living together. Living things tend to live on or in other living things, and such associations as these may be the most common form of life on earth.

What is the result? A beautiful blue-stained wood called blue stain pine. Have you seen it? You may have seen it as rustic siding and paneling on cabins or homes, or in mountain dulcimers made in Colorado. I find it fascinating the way nature creates beauty through natural processes that sometimes seem ugly and harsh.

"Everything is dangerous, and one must decide whether to live in fear, or courage"
Emma Hamm,

BEAUTY AND BEASTS: poem

WHEN BEAUTY IS THE BEAST

When beauty is the beast
Death seems a delicate thing
Surrounded by iridescent colors
It arrives on transparent wings

Then the air hangs heavy
And the sun lifts light
Sparkling off the water
The blue beauty alights

Death folds her wings and waits
For the music only she can hear
To begin her deadly dance
That the careless fear

At length dancing with grace
Death hovers through the sky
Shimmering in the sun
Bloodthirsty damselfly

BEAUTY AND BEASTS: science

Scientists often say their theories must be beautiful. Do artists think their art must be true?

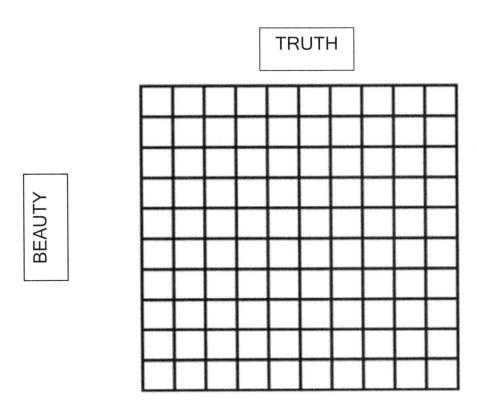

Art often portrays ugly, provoking subjects. Is science always benign?

BEAUTY AND BEASTS: art - draw - paint - paste - fold - etc.

Chapter 19

EPILOGUE

"Many that live deserve death.
And some that die deserve life.
Can you give it to them?
Then do not be too eager to
deal out death in judgement.
For even the very wise
cannot see all ends."
J. R. R. Tolkien

COMBINATORIAL FLUENCY

There are several problems involved in trying to become a scientist. Not the least of these is the social stigma, but I can't deal with that in this space. I can hardly deal with it in other spaces. I also don't want to make becoming a scientist sound too hard because I have found that scientists never get any sympathy for our plight anyway.

We do have a plight, however, which is not the same as having a fit. The latter is a medical condition. As far as I can tell, the difficulty lies in the fact that doing science has almost nothing to do with becoming a scientist. This may, at first, seem strange to you. However, I suspect it is a similar problem to many other activities.

For example, I can sometimes write a complete sentence, and once I wrote a compelling newspaper column. However, to gain this skill I had to learn how to diagram sentences in the eighth grade. I was told that this was an imperative skill and that Shakespeare could never have created his art without such mastery.

I can truthfully testify to you that I have never diagramed a single sentence of a single column in this newspaper, my dissertations, or any of my scientific articles. My wife, who is my first editor, says she can tell, but then she is an obsessive ex-schoolteacher. So becoming a writer is different than being a writer.

Shakespeare apparently had what I call "combinatorial fluency" with words (which term I stole from someone else more fluent than I, but I can't find the book right now to tell you who it was). Combinatorial fluency is the ability to utilize whatever jargon or code you are dealing with effortlessly and fluently, so that the elements can be combined in new and creative ways.

Learning facts and figures does not make one a scientist. The elements of the given discipline must be learned in such a way as to achieve combinatorial fluency: the mastery of the ideas of a discipline so completely that it allows you to recombine ideas into new insights. This is frequently done through language, although the language sometimes looks like math or strange chemical symbols. Sometimes the language looks like computer code.

One of the skills that allows creative new use of scientific ideas is the computer. However, the computer has its own firm code that requires fluency as much as any language if it is to be used creatively. (Get it? Firm code? . . . Sorry!) At one time I was encouraged by the use of technology by the younger generations. They are born into technology and its use is everywhere. They know how to navigate the latest phone, laptop, tablet, and desktop.

However, using computer technology has almost nothing to do with being a computer programmer. While the youth are native users and technologically proficient, they are not, for the most part, combinatorial fluent. They are technologically fluent. In fact, many cannot write computer code at all. They may spend countless hours using technology, but most have little knowledge of how it works.

Of course there is a certain stigma to being a computer scientist. I can't deal with that here. But I can tell you that learning to code is an essential step for advancing creativity and science in this modern world. Like becoming a scientist, it isn't all that hard. There are now numerous online sources that teach people how to code.

As far as I can tell, as an amateur programmer, using technology has almost nothing to do with programming technology. The numerous computer codes are just another language. If one wants to write the equivalence of a Shakespearean classic in it, you will have to develop combinatorial fluency. I suspect that those who fail to do so in the coming years will not receive much sympathy for their plight.

One of the problems facing todays science is that in mastering the language of communication the culture the inhabit. Then they must master the language of their field of specialization to the point of combinatorial fluency. This often leaves little time for becoming combinatorial fluent in the scientific method of thinking and experimenting.

No poems

No thoughts

No coloring pages

No science

No whatever's

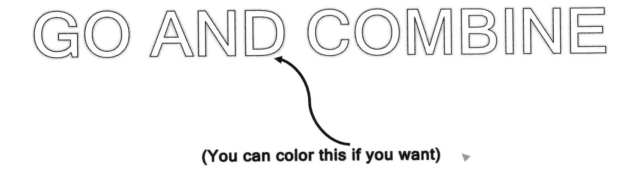

(You can color this if you want)

ABOUT THE AUTHOR

Dr. Gary L. McCallister is Professor Emeritus from Colorado Mesa University. That means he no longer professes anything at said university. He professed things biological at said university for over forty years. He is a highly trained professional windbag and can prove it with sixty-two scientific papers, eight books (so far), and an award-winning weekly science column for the newspaper. He has also produced 15 music CD's and is a luthier of the mountain dulcimer. His unique designs are popular throughout the western United States. He lives in western Colorado where he tends bees, grows prickly pear cactus, and plays music on his guitars, mandolins, banjos, and mountain dulcimers. You can contact him at: gmccallister@bresnan.net

OTHER BOOKS BY GARY MCCALLISTER

MUSIC
Making More than Music 2014
First Songs with the Mountain Dulcimer: history, instrument, and simple songs 2015
Hymns on Mountain Dulcimer: Learn to play the mountain dulcimer using hymns 2016

SCIENCE
Hanging Out With GRAVITY: Galileo's gravity game 2015
Seriously Silly Science: A science reader for the whole year – and some of it is even true 2015
A Convenient Truce: A cease fire in the war between religion and science 2016
The Solar Solution: the solution to problems you didn't even know you had. 2017
Between Two Mirrors: art and science in the modern world 2017
Science is Serous: all the scientists say so 2018
Thou Shalt Make: the spiritual significance of making things 2019
Reality: the cure for poor behavior 2019
A Simply Scientific Christmas: Holiday Talk for the Dinner Table (Simply Scientific Holiday series Book 1) 2020
A Simply Scientific" Thanksgiving: Gratitude Measured in Metrics ("Simply Scientific" Holidays Book 2) 2020
A Simply Scientific" Valentines: Romance for Nerds (Simply Scientific Holiday Series Book 3)
A Simply Scientific" Valentines: Romance for Nerds (Simply Scientific Holiday Series Book 3) 2020
A "Simply Scientific" 4th of JULY: The Science of Freedom (Holiday Science) 2020

HISTORY
Walking Man 2015
The Hammars of History 2018 (Editor and publisher)
The Book of Lehi 2020

All available on Amazon.com